The Best American Science Writing 2006

THE BEST AMERICAN SCIENCE WRITING

EDITORS

2000: James Gleick
2001: Timothy Ferris
2002: Matt Ridley
2003: Oliver Sacks
2004: Dava Sobel
2005: Alan Lightman

The Best American Science Writing 2006

Editor: Atul Gawande

Series Editor: Jesse Cohen

AN ecco BOOK

HARPER PERENNIAL

NEW YORK • LONDON • TORONTO • SYDNEY

HARPER PERENNIAL

Permissions appear following page 362.

FIRST EDITION

Designed by Cassandra J. Pappas

Library of Congress Cataloging-in-Publication Data is available upon request.

ISBN-10: 0-06-072644-X
ISBN-13: 978-0-06-072644-7

06 07 08 09 BVG/RRD 10 9 8 7 6 5 4 3

Contents

Introduction by Atul Gawande

THE STATE OF KANSAS, until its creationist board of education stepped in last year to muddy the pond, had a rather useful definition of what science is. "Science," the official definition said, "is the human activity of seeking natural explanations for what we observe in the world around us." There is still some disagreement among scientists about what that human activity must be like to qualify as science. But for the most part, if you search for natural explanations of reality systematically, using theories of causation that can be tested through observation and experimentation, everyone will agree: you're doing science.

But when are you doing science writing? The answers get pretty squishy. My own definition would be this: Science writing is writing about the scientific investigation of the world, about the knowledge acquired, or about what happens to that knowledge when it is thrown back into the world. I think that about covers all the bases.

So then what would the definition of the best science writing be? The clearest, most completely objective answer is: the best science writing is science writing that is cool. Even better, this particular year the best science writing is science writing that *I* think is cool. And there are all kinds of science writing that I think is cool.

I like science writing to be clear and to be interesting to scientists and nonscientists alike. I like it to be smart. I like it, every once in a while, to be funny. I like science writing to have a beginning, middle, and end—to tell a story whenever possible.

Among the essays that series editor Jesse Cohen had carefully culled for my consideration, there were three that I loved simply for the story they told. I have no inherent interest in chess programs, or how the brain processes music differently from speech, or the occurrence of progressive supranuclear palsy in Guam. But damn if the writers on these obscurities didn't manage to tell a thrilling story anyway. Tom Mueller's "Your Move" unfolds the story of Chrilly Donninger, an obsessive Austrian who was hired by an Arab sheikh to create the best chess program in the world and in the process produced a program so advanced that it is creating chess strategies human beings have never seen before. In "My Bionic Quest for *Boléro*," Michael Chorost, who went completely deaf at age thirty-seven, describes his methodical effort to reengineer the software in his cochlear implant so that he might hear Ravel's masterpiece again. Jonathan Weiner, in his scientific detective story "The Tangle," follows an ethnobotanist (whatever that is) as he investigates an epidemic of a strange neurologic disease and finds answers that might explain Lou Gehrig's disease. The stories have characters and twists. Most of all, though, the writers show a feel not just for the drama of the human tale but also for the drama in the ideas themselves.

Others of the essays here stood out because they reveal something unexpected about a province of our world that we thought we understood. W. Wayt Gibbs's essay on obesity calmly and lucidly demolishes our received wisdom that obesity kills hundreds of thousands of Americans a year. (This is particularly galling for me, since I'm one of those who have blindly quoted as fact the statistics he eviscerates.) Kenneth Chang, in "Ten Planets? Why Not Eleven?" romps with almost evil delight through the confusion over whether the recent discovery of an orbiting ice ball larger than Pluto marks the discovery of a tenth planet or the demotion of Pluto to an asteroid. Paul

Bloom's "Is God an Accident?" ponders whether the ideas of a soul, the afterlife, and God himself could arise from our inherent but mistaken tendency to believe that the mind and the body are separable entities. Both Robert Provine and Dennis Overbye examine things that are similarly everyday and ordinary—yawning in Provine's case, and time in Overbye's—and carry us along enthralled as they show the fascinating possibilities inside. Science writing can show the complexity in the most seemingly simple of phenomena.

There are other essays here, however—some of the most important—that manage to do almost the opposite. They examine areas of science in which there is great public bewilderment and discord—sometimes dangerous discord—and pierce that confusion with shining clarity. H. Allen Orr's "Devolution" was the first magazine piece to comprehensively dissect the claims of intelligent design theorists. Elizabeth Kolbert's epic three-part series in *The New Yorker*, "The Climate of Man," from which "The Curse of Akkad" was selected, examines everything from climate modeling to glacier and ocean temperature measurements to data from archaeological finds to consider the arguments for taking global warming seriously. Gardiner Harris and Anahad O'Connor's disturbing article on whether vaccines containing mercury cause autism or not sounds a serious alarm for scientists about the depth of public mistrust in our authority and explanations. Neil Swidey's careful piece on "What Makes People Gay?" takes us through a political minefield of scientific studies never forgetting that human beings are the subject of his investigation. This is science writing as public service.

Another class of science writing might be called "nuts and bolts stories," stories that reveal the scientific process itself in all its uncertainty and human complexity. Jonathan Weiner's "The Tangle" would also fall in this category, and so would Michael Specter's timely, masterly, and sobering piece, "Nature's Bioterrorist," on avian influenza and exactly how scientists are going about trying to stop a deadly pandemic. In "The Day Everything Died," Karen Wright brings us along to see the work of Luann Becker, a young geologist with a controversial theory, and gives us a gripping inside look at the scientific battle over what caused the Permian extinction 250 million years ago, which wiped out more than 90 percent of marine species. Jack Hitt's "Mighty White of You," about the battle to determine who the first Americans were, gets us so close to the science of archaeology, so far inside the massive ambiguities in the data, we begin to wonder whether archaeology is a science at all.

Some pieces here accomplish an altogether different task. They show us how science might illuminate areas of life that we don't commonly bring science to help with. The most obvious example is D. T. Max's "Literary Darwinists," an essay in which Max takes us through the sometimes successful—and sometimes less than successful—fledgling efforts of theorists to understand literature using evolutionary psychology. "The Coming Death Shortage" by Charles Mann is a scientific polemic—and a depressingly compelling one at that—on the now fragile structure of human society. It marshals evidence from science to illuminate the larger question of what science has wrought upon society by doubling our millennia-established longevity in the past several decades. In "Earth Without People," Alan Weisman performs a fascinating thought experiment, considering what would happen to our planet if people suddenly disappeared. Frans B. M. de Waal's brilliant concluding essay, "We're All Machiavellians," asks why science has had so little to say about the nature and importance of the drive for power in human beings. One would have thought scientific investigation and explanation had already penetrated all possible corners of the natural world. But with great creativity, these authors show it is hardly the case.

Finally, the reader will find at least one essay whose selection I can't completely explain. A few years ago in his long profile of the maverick human genome decoder J. Craig Venter, Richard Preston stopped and took a moment to describe what a small aliquot of his own purified DNA tasted like when he dropped the clear sticky goo onto his tongue. It was an indelible description. There was nothing really scientific about it. He had no theory or hypothesis he was testing—no evident purpose and certainly no objective method of measurement. It was purely subjective. Yet he captured in that moment a deeply scientific impulse. And he does it again in "Climbing the Redwoods," his essay included here. He describes climbing with a scientist into the dense canopy at the intertwined top of a cluster of giant redwood trees, its own living world three hundred feet off the ground, and his detail is the marvel. He finds lichen growing on the bark "like tiny pumpkin pies," a fire cave, a crisscrossing of branches so thick there were "fusions, bridges, and spires," and he could not see the ground.

Is this science writing? Maybe. It certainly is cool.

The Best American Science Writing 2006

TOM MUELLER

Your Move

FROM *THE NEW YORKER*

When the IBM supercomputer Deep Blue defeated chess champion Garry Kasparov in 1997, it was heralded as a major turning point in the continuing struggle between man and machine. Adapting Deep Blue's approach to less powerful PCs, programmers are making up for lack of number-crunching ability with artfulness. As Tom Mueller has discovered, the results have been unexpected, with computers developing strategies grand masters have never thought of.

Chrilly Donninger prefers to watch from a distance when Hydra, his computer chess program, competes, because he is camera-shy, but also because he rarely understands what Hydra is doing, and the uncertainty makes him nervous. During Hydra's match against the world's seventh-ranked player, Michael Adams, in London last June, Donninger sat with three grand masters at the back of a darkened auditorium, watching a video projection of the competition on the wall behind Adams. Most of the time, Donninger, a forty-nine-year-old Austrian, had little to worry about; Hydra won the match five games to none, with one draw. But in the second game, which ended in the draw, the program made an error that briefly gave its human opponent an advantage.

The game was played at a spotlit table on a low podium. Adams sat in the classic chess player's pose—his elbows resting on the table, his chin cupped in his palms—reaching out now and then with his right hand to move a piece on a large wooden chessboard. Across from him was Hydra—a laptop linked by Internet connection to a thirty-two-processor Linux cluster in Abu Dhabi—and Hydra's human operator, who entered Adams's moves into the computer and recorded the program's replies on the board. On the laptop's screen was a virtual chessboard showing the current position in the game, as well as a pane of swiftly scrolling numbers representing a fraction of the thousands of lines of play that Hydra was analyzing, and a row of colored bars that grew or shrank with each move, according to the program's assessment of who was winning—green bars meant an advantage for white, red bars for black.

For much of the match, the bars showed Hydra comfortably in the lead. When Adams made a mistake, they spiked dramatically, but mostly they grew in small increments, recording the tiny advantages that the program was steadily accumulating. Many of these were so subtle that Donninger and the grand masters failed to grasp the logic of Hydra's moves until long after they had been made. But about twenty minutes into the second game, when Hydra advanced its central e-pawn to the fifth rank, there was a small commotion in the group. Yasser Seirawan, an American player formerly ranked in the top ten, who had coached Adams for the match, gave a thumbs-up sign. Christopher Lutz, a German grand master who is Hydra's main chess adviser, groaned. Only Donninger, who programs chess far better than he plays it, was baffled. He turned to Lutz in alarm.

"What was that? What did you see?"

"Now our pawn structure has become inflexible," Lutz replied. "Do we have anything in the program for flexibility?"

"What do you mean by 'flexibility'?"

Lutz frowned. He sensed that Hydra had hemmed itself in, giving Adams the upper hand. Bishop to b7 was the correct move, Lutz believed—the most natural way for Hydra to preserve its attacking chances and its room to maneuver. But explaining his nebulous insights to a lesser player like Donninger was a challenge.

"This position lacks flexibility," he repeated, shaking his head.

"When you can define 'flexibility' in twelve bits, it'll go in Hydra," Donninger told him, twelve bits being the size of the program's data tables.

Adams locked up Hydra's center with his next move and managed, several hours later, to eke out a draw. "Hydra didn't play badly, but 'not bad' isn't good enough against a leading grand master," Donninger said after the game. His program is widely considered to be the world's strongest chess player, human or digital, but it still has room for improvement.

LEAN AND RESTLESS, with a scraggly beard and a large Roman nose, Donninger says that he approaches programming less like a scientist than like a craftsman—he compares himself to a *Madonnenschnitzer,* one of the painstaking Baroque and rococo wood-carvers whose Madonna sculptures adorn the churches near Altmelon, the village in northern Austria where he lives and works. He speaks German with a thick Austrian brogue and frequently uses expressions like *"Das ist mir Wurscht!"*—"That's all sausage to me!" For the past two years, he has led the Hydra project, a multinational team of computer and chess experts, which is funded by the Pal Group, a company based in the United Arab Emirates which makes computer systems, desalinization plants, and cyber cafes. Pal's owner, Sheikh Tahnoon bin Zayed al-Nahyan, is a member of the country's royal family and a passionate chess player; he hired Donninger with the goal of creating the world's best chess program. Pal is also using the same kind of hardware that runs Hydra for fingerprint-matching and DNA-analysis applications, which, like computer chess, require high-speed calculations. The program's main hardware resides in an air-conditioned room in Abu Dhabi, and Donninger is frequently unable to access it, because the sheikh and Hydra, playing under the name zor_champ, are on the Internet, taking on all comers.

As a child, Donninger was so attached to puzzles that his mother worried that he was disturbed. At the age of four, he spent months building houses out of four colors of Lego bricks, in which no bricks of the same color ever touched; two decades later, when he was an undergraduate at the University of Vienna, he learned that this was a famous conundrum in topology—the Four-Color Problem. After completing a doctorate in statistics, he worked as a programmer for Siemens, where he earned a reputation as a bug fixer, the computer equivalent of a puzzler. In 1989, he was transferred to the Dutch city of Noordwijk. It was there, during a period of intense loneliness, that Donninger joined a local chess club and started writing his first chess program. "I found my ecological niche," he says.

He had also found the ultimate puzzle. With about 10^{128} possible unique games—vastly more than there are atoms in the known universe—chess is one of mankind's most complex activities. In an average arrangement on the board, white has thirty-five possible moves and black has thirty-five possible replies, yielding twelve hundred and twenty-five potential positions after one full turn. With subsequent moves, each of these positions branches out exponentially in further lines of play—1.5 million positions after the second turn, 1.8 billion after the third—forming a gigantic map of potential games that programmers call the "search tree."

How human beings confront this complexity and seize on a few good moves remains a mystery. Experienced players rely on subconscious faculties known variously as pattern recognition, visualization, and aesthetic sense. All are forms of educated guesswork—aids to making choices when certainty through exhaustive calculation is impossible—and may be summed up in a word: intuition. Even a novice player uses intuition to exclude most moves as pointless, and the more advanced a player becomes the less he needs to calculate. As the eminent Cuban grand master José Raúl Capablanca once told a weaker player, "You figure it out, I know it."

Computers have the advantage of formidable analytical power, but even the fastest machine is quickly overwhelmed by the sheer number of moves that it must assess. (Donninger estimates that Hydra would need 10^{30} years to "solve" chess, starting from the first move and analyzing all possible sequences of play.) To produce world-class chess of the sort that Hydra played against Michael Adams, programmers must somehow teach their machines intuition.

This turns out to be a highly personal task, which every programmer approaches differently. Stefan Meyer-Kahlen, a thirty-seven-year-old German who wrote the four-time world-champion program Shredder, was inspired by Anatoly Karpov, the Russian player known for his calm, ruthlessly logical play and his masterly defense. Amir Ban and Shay Bushinsky, Israeli programmers who created Junior, another four-time world champion, draw on a large collection of computer chess games to shape their program's style. "We don't use grand-master games for this, because they're too full of errors," says Ban, who made a fortune in the 1990s as an entrepreneur in flash memory. (Bushinsky is a professor of computer science and artificial intelligence, or A.I., at the University of Haifa.) Ban and Bushinsky believe that their method accounts for Junior's "speculative" play—its keen understanding of time and space, and its unrivaled knack for sacrifice—and they gleefully describe how Junior trounced an early version of Hydra two years ago, because Hydra (or, rather, Donninger) had badly misjudged a position. (Ban and Donninger detest one another, and after getting into a shouting match at a tournament in 1997, they no longer speak.)

Hydra is famous for its relentless assault on the enemy king. "It's the Rottweiler of computer chess," Donninger says proudly. "It floats like a butterfly, stings like a hornet." He says that he has tried to endow the program with the slashing, sacrificial style of the former world champion Mikhail Tal. Yet Hydra's fighting spirit is as much Donninger's as Tal's. As a boy, Donninger fought constantly in school, analyzed each of Muhammad Ali's matches, and taught his younger sister an Ali-inspired uppercut to ward off bullies. In his twenties, he fought as a junior welterweight in a top Vienna gym. Occasionally, he pulls on a worn pair of boxing gloves and hammers away at a heavy bag that hangs from a beam on the side of his house.

Hydra, Shredder, Junior, and Fritz—another top program—routinely defeat leading grand masters, each playing with a distinctive personality that reflects not only its inventor's particular approach to chess and programming, but also moves and tactics that seem to arise spontaneously from intricacies of the computer code, which the programmer himself often cannot explain. Over the past decade, these programs, which typically sell for about sixty dollars and run on a PC, have become essential tools for grand masters to analyze past games, test opening lines, and generate new ideas. (Hydra is not commercially available, in part because of its specialized hardware; some programs, including versions of the reigning world champion, Zappa, can be downloaded, free, from the Internet.) In the past several years, these programs have begun to play the kind of elegant, creative chess once thought to be the exclusive province of humans. Computers are allowing people to look more deeply into the game than ever before. They are even helping people to play more like machines.

The first chess automaton was, like Hydra, created as a diversion for royalty. In 1769, the engineer Wolfgang von Kempelen built "the Turk" for the Hapsburg ruler Maria Theresa. A robed and turbaned mannequin seated at a large desk, the Turk toured Europe and America for decades, trouncing all but the best players; according to one story, it beat Napoleon so badly that he swept the pieces from the board in disgust. Though the Turk's mechanical components were impressive—it had elaborate, whirring gears and could nod, roll its eyes, and lean over the board as if in concentration—its human component was decisive: it was secretly operated by a skilled chess player curled up inside the desk.

The first real chess-playing machine was designed in the late 1940s by the British mathematician Alan Turing, who, in 1950, in the paper in which he proposed the famous Turing Test, identified chess as an ideal proving ground for machine intelligence. Since computers were not yet widely available, Turing acted as his own central processing unit, laboriously working out each move on paper. With the advent of computers a few years later, chess became the darling of the A.I. community. The game was enormously complex, had simple rules, and, unlike many objects of scientific research, provided an unambiguous way to measure results: the more a program won, the smarter it was thought to be. In broad terms, the early programs of the 1950s and 1960s worked the way Hydra and its peers do today. They consisted of a search function, which sifted through the tree of possible moves, and an evaluation function, which applied a score to positions that arose along the way, awarding bonuses for good characteristics and penalties for bad. The move leading to the highest-scoring position would be selected.

The first chess programmers attempted to use the same intuitive strategies in search and evaluation that people employed when playing chess. However, their translations of human rules of thumb—control the center, don't move your queen too early—into programming language proved too crude to produce good chess. Worse, the more knowledge the programmers built into the search function, the slower the search became, which limited how deeply the program could see into a position—its "search horizon." The first programs were characterized by nonexistent strategy, embarrassing endgame technique, and a remarkable gift for blunders. In 1956, the MANIAC, a nuclear-weapons computer at Los Alamos, lost to a human opponent, even though the rules had been simplified and it had been spotted a queen. Mikhail Botvinnik, a former world champion, tried to write a program that thought the way he did, but it never won a game in tournament play.

A breakthrough came in the late 1970s, with the advent of "brute force" programming, which traded selectivity for speed. By paring back chess knowledge, emphasizing search algorithms like alpha-beta pruning, and exploiting faster hardware, programmers were able to consider a far broader search tree. Computers began to beat master players in the early eighties, and in 1988 Deep Thought became the first program to defeat a grand master. The brute-force approach culminated in 1997, when Deep Thought's successor, Deep Blue, a multimillion-dollar IBM supercomputer that could evaluate two hundred million positions per second, won a six-game match against Garry Kasparov, the world champion.

Computers had triumphed at chess not by aping human thought, as most A.I. experts had expected, but by playing like machines. The analogy with flight is instructive: as long as people tried to fly by imitating birds, attaching diaphanous wings to their arms and flapping madly, they were doomed to failure; once they escaped the paradigm of the familiar, however, they were soon flying much faster than birds. Yet something still seemed to be missing. Programmers had made the most of computers' computational superiority, but many experts agreed that humans retained the edge in strategy. After Deep Blue's victory, A.I. researchers lost interest in chess, largely because brute-force methods seemed too crude and mechanical to shed much light on the nature of intelligence. " 'Brute force' was a derogatory term," recalls Jonathan Schaeffer, a computer scientist who is the author of several chess programs and the world-champion checkers program Chinook. "You were considered a heretic if you didn't try to emulate the human brain."

By the late nineties, most chess programming was done not for mainframes but for much slower microcomputers and, subsequently, for PCs. Even today, the fastest PC programs can analyze only about four million positions per second, a fraction of Deep Blue's capacity. In order to achieve world-class results at these speeds, programmers needed to find ways of searching more selectively and evaluating more precisely. Over the past several years, the most gifted programmers have learned how to distill the more arcane principles of grand-master play into computer language. Some have identified principles that grand masters never imagined.

Donninger reviews Hydra's errors alone or with Christopher Lutz, trying to define each problematic position in twelve bits' worth of questions, which he can incorporate into Hydra's code. "Sometimes it would be nice to have more questions to work with," Donninger says. "But then you'd risk diluting the essential characteristics of the position with less vital information. Twelve bits is good discipline." Because these questions determine how Hydra will play, Donninger won't reveal most of them; a few, however, are more or less the same for all programs. The questions that Hydra uses to assess the value of a passed pawn are: "Is this a passed pawn?" (1 bit); "Is the square in front of it blocked?" (1 bit); "Which row is it on?" (3 bits); "Is it supported by a neighboring pawn?" (1 bit); "Is it supported by a neighboring pawn that is itself a passed pawn?" (1 bit); "Is it an advanced passed pawn?" (1 bit); "Is the enemy king inside the pawn's square?" (1 bit); and "Which phase of the game are we in?" (3 bits).

Once Donninger is satisfied that he has the right questions, he enters them into Hydra's evaluation function as a new heuristic, or chess rule. Hydra, like all other chess software, has hundreds of heuristics woven into its code, where they behave like DNA, shaping the program's personality. The Hydra code for evaluating a passed pawn, written in the computer language Verilog, looks like this:

```
if(iwPA7) begin
    wBlocked_A<=!iEmptyA8;
    wPassed_A<=1'b1;
    wSupported_A<=iwPB6;
    wDuo_A<=(iwPB7);
    wBack_A<=1'b0;
    wMaxRow_A<=3'h6;
    wPRow_A<=3'h6;
    wLever_A<=1'b0;
end
```

How the heuristics interact, reinforcing and overriding one another, is mysterious; even a slight adjustment to a single rule can produce side effects that the programmer cannot predict. After Donninger adds a new rule, he runs a long series of test matches to determine whether it works, pitting the new version of Hydra against other programs. If the new Hydra wins more often than its predecessor, the new heuristic stays. Donninger says that for each new rule he needs about three months to work out unexpected kinks. (By contrast, Ban and Bushinsky sometimes tweak Junior's code right up to the start of a game. "We like to change things, take risks, improvise," Bushinsky says. "Maybe this is not so smart sometimes—it's considered a real no-no in computer science. But that's how we work. Maybe we do have an instinct for the program, sense something about how it feels.")

Today, the best programs blend knowledge and speed so effectively that even the most talented human players have little chance of defeating them. In 2003, Fritz and Junior fought Garry Kasparov to a draw in tense multi-game matches. A round-robin event in October, 2004, pitting Hydra, Fritz, and Junior against three leading grand masters, ended in an 8½–3½ victory for the computers. The machines' superiority was most obvious during the Hydra-Adams match in London. "Adams was simply pushed off the board by a much stronger opponent," says David Levy, an international master who watched the event.

Donninger is no longer interested in man-versus-machine matches. "I see the same pattern in each game," he says. "I call it Chrilly's Law: every ten moves, at the most, in complicated positions, even the strongest player will commit a slight inaccuracy—the second-best move when only the best will do. He doesn't even notice it, but Hydra does. Its evaluation bars start growing, a little taller with every move. By the time the grand master realizes the problem, it's already Game Over." Many chess players and programmers would like to see a match between Hydra and Kasparov, who has officially retired from chess to pursue a career as a politician in Russia but is still considered the ultimate opponent. "The world's greatest-ever human player against the world's greatest-ever computer player—we would all love to see it," Levy says. Donninger is indifferent to the idea. "I'm much more interested in beating Shredder, Fritz, and the other programs," he says. "I learn more from those matches."

IN FEBRUARY, I watched Hydra play Shredder in one of the most sophisticated chess games in history, during the fourteenth International Computer Chess Championship. The tournament was held in Paderborn, Germany, in a shabby, fluorescent-lit conference room whose boisterous disarray often suggested a tailgate party more than a competition. Sixteen contestants sat chatting in pairs at small tables laden with computers, chess sets, beer, coffee, and half-eaten hunks of strudel. The programs reflected their authors' whims: one sounded a gong with every move, others displayed photos—of a pet falcon in one case, a scantily clad starlet in another. "They're extensions of our egos," Vincent Diepeveen, the author of Diep, a Dutch program, said.

Donninger and Hydra sat at a table in the middle of the room, the laptop draped with a UAE flag and Donninger in his game-day outfit, which he wore on all five days of the tournament: black jeans, dirty white clogs, and a gray cardigan with deerhorn buttons. Across from him was Shredder and its inventor, Meyer-Kahlen, a round-faced man with a melodious tenor voice, who is the only professional chess programmer not affiliated with a larger organization and is widely admired by his peers. "He does everything himself—the program, the user interface, sales, technical support," Donninger says. "No one knows how he manages to stay at the top."

The game started with a handshake, a custom in matches between grand masters, and continued for about twenty minutes with a chatty congeniality unthinkable in competitive chess. Once a game begins, programmers are not permitted to adjust their programs, but there is no rule against sharing information. Donninger and Meyer-Kahlen turned their screens so that each could see from the scrolling numbers and colored bars what the other's program was thinking, and speculated aloud about their programs' prospects. When Donninger recorded the wrong move on the board, Meyer-Kahlen politely corrected it. However, as the position became complicated, both men grew quiet and stone-faced, their eyes fixed on the screens. Other programmers drifted away from their matches and gathered in a tight circle around the table.

"What a stupid sport!" Donninger snapped, as the programs jockeyed for infinitesimal advantages. "We're completely helpless. It's like riding shotgun in a Formula One race car."

Ninety minutes into the game, most of the pieces were still on the board, arranged in an intricate logjam. Shredder was happy with the position, but so was Hydra; evidently one of the programs was mistaken. Then Shredder attacked Hydra's kingside, with a risky pawn advance that seemed to expose Shredder's own king. "*Scheisse!*" Meyer-Kahlen exclaimed. "What the devil is Shredder doing?"

The situation was critical, but neither programmer knew who was winning. Nor did anyone in the crowd around the table, which included the two grand masters on the Hydra team, Christopher Lutz and Talib Mousa.

"*Unklar,*" Lutz murmured, when I asked him what he thought of the position.

Donninger's face brightened for a moment. " 'Unclear' is a grand master's way of saying, 'Who the hell knows?' "

The uncertainty persisted for several more moves. Then, suddenly, it became obvious that Shredder was on the verge of being checkmated. Nevertheless, the program made a queenside pawn push, just as Hydra was cornering its king on the other side of the board.

"Oh, Shredder, what kind of a crap move is that?" Meyer-Kahlen said. Eight moves later, he reached abruptly across the board to shake Donninger's hand, resigning the game. Then he stalked away from the table.

Afterward, while Mousa recounted the match to Sheikh Tahnoon on a cell phone, Lutz and Donninger reviewed the game at the chessboard. "We're trying to understand what happened and why," Donninger said wearily.

It is now plausible to argue that computers are playing subtler, more imaginative chess than the humans they have been designed to emulate. "They make a lot of counterintuitive, even absurd-looking, moves that on closer inspection can turn out to be outrageously creative," says John Watson, an international master who has written more than twenty-five books on advanced chess theory and strategy. "By generating countless new ideas, they are expanding the boundaries of chess, enabling top players to study the game more deeply, play more subtly." Viswanathan Anand, a thirty-five-year-old Indian who is currently rated the world's top active player, often uses several chess programs simultaneously when he trains. "I have Shredder, Fritz, Junior, and HIARCS"—another popular program—"running all the time so I can see their various opinions, which are often very different," he says. "When a position catches my fancy, we compare notes." In some cases, computers are rewriting the game's ground rules. "Certain endgames that for centuries were unanimously thought to be draws have actually proved to be clear wins," says John Nunn, a player formerly ranked in the top ten, whose groundbreaking books on endgames were the first to be written using extensive computer analysis. "Computers helped me to discover a number of fundamentally new positions that no one had ever expected, some of which were outstandingly beautiful," he says.

Chess programs are even having a psychological impact. Because computers feel neither nervousness nor fear, they are able to defend apparently hopeless positions, which has encouraged human opponents to persevere even when defeat seems inevitable. The seventeen-year-old American champion Hikaru Nakamura recently told the *Times* that he plays with the courage of a computer. "I'll play some of these really crazy moves that people are not going to be expecting," he said. "The way I play is not like most people. The moves are more computeresque."

Shay Bushinsky and Amir Ban believe that computeresque is better. "Many people don't like it when I say this, but I think Junior plays more creatively than humans," Bushinsky says. Ban goes further, insisting that Junior's creativity is a symptom of its inherent intelligence. "This is an emergent phenomenon of the program, not something I put into it," he says. "It's like Junior is the child and I'm the father. I may think I've taught my child everything, but it's constantly dreaming up things that surprise me."

Such claims to creativity and intelligence are not that far-fetched. In the last few decades, scholars of emergent phenomena have revealed how simple rules at work in termite mounds, traffic jams, quantum mechanics, and the structure of galaxies can give rise to sophisticated and unanticipated behavior, just as the few basic rules of chess yield the endlessly subtle game played by grand masters and chess programs. Scientists have found that the best way to understand these complex systems is often not to theorize but to build a computer model and see how it behaves, in the same way that a chess programmer writes a new version of his program and then watches it play. As the model exhibits patterns and behaviors whose existence its programmer never suspected, it is, in some real sense, creating. And while this may not constitute true machine intelligence, a growing number of cognitive scientists and philosophers see no fundamental distinction between computers and human brains. "Sure, I think the brain is a machine," Mark Greenberg, a philosopher of mind at UCLA, says. "And, likewise, that there's no reason in principle that a computer couldn't think, have beliefs and other mental states, be intelligent. Many mainstream philosophers would agree."

Some, of course, do not. John Searle has tried to disprove the notion of machine intelligence with his "Chinese Room" thought experiment, and Roger Penrose has used Gödel's ideas to the same end. The philosopher Colin McGinn argues that consciousness itself, a crucial part of intelligence, is cognitively closed to us. Yet for those who view intelligence as something essentially and almost mystically human, the analogy with flight may provide a salutary warning. In October 1903, just two months before the Wright brothers completed their first successful powered flight, the astronomer Simon Newcomb published an essay attempting to prove that airplanes would never fly. He got nearly everything right, identifying the intricate ratios of size, weight, and wing surface in birds, which proved that a man-size bird could never get airborne. Yet he overlooked one key detail—the lift effect of an airfoil—and the larger point that an airplane was not a bird. Might those who appear to require a human brain for intelligence be overlooking a broader but no less valid definition of the term?

In any event, such metaphysical matters can be neatly sidestepped with the Turing Test, which merely asks whether a machine can imitate intelligence well enough to fool a human observer. In 2000 and 2001, the leading British grand master Nigel Short played a number of speed-chess games on the Internet, against an anonymous opponent. In most games, this player put himself at a disadvantage with several bizarre opening moves, yet went on to trounce Short, who is among the world's best speed-chess players. Short became convinced that his quirky, brilliant opponent was the reclusive chess genius Bobby Fischer—who else could beat him with such superhuman ease? "I am ninety-nine percent sure that I have been playing against the chess legend," he told the London *Sunday Telegraph.* "It's tremendously exciting." He said that he treasured these games as products of Fischer's rare art. "To me, they are what an undiscovered Mozart symphony would be to a music lover."

In fact, Short's opponent was probably a computer. "It's fairly clear that the phantom Fischer was an experienced chess-program user playing a practical joke," Frederic Friedel, a founder of ChessBase, a software company that publishes Fritz, Junior, and several other programs, says. "He made the first few absurd moves by hand to throw people off the scent, then unleashed the machine." When Friedel played through several of the games with Fritz, the program's moves were virtually identical to those of Short's mystery opponent. So a computer duped one of the world's top chess players into believing that it was a human. Friedel says that the converse also happens. "Raffael," one of the strongest players at his company's popular chess server Playchess.com, was originally thought to be a powerful computer, but was later observed making all-too-human errors. "Raffael" is now believed by many to be Garry Kasparov.

MICHAEL CHOROST

My Bionic Quest for Boléro

FROM *WIRED*

When Michael Chorost lost his hearing, he was fitted with a computer—a cochlear implant—surgically placed in his head to stimulate his auditory nerves. The effect was immediate, but cochlear implants are optimized for speech. To hear music, especially Boléro, *he embarked on a search for new software for his internal computer.*

With one listen, I was hooked. I was a fifteen-year-old suburban New Jersey nerd, racked with teenage lust but too timid to ask for a date. When I came across *Boléro* among the LPs in my parents' record collection, I put it on the turntable. It hit me like a neural thunderstorm, titanic and glorious, each cycle building to a climax and waiting but a beat before launching into the next.

I had no idea back then of *Boléro*'s reputation as one of the most famous orchestral recordings in the world. When it was first performed at the Paris Opera in 1928, the fifteen-minute composition stunned the audience. Of the French composer, Maurice Ravel, a woman in attendance reportedly cried out, "He's mad . . . he's mad!" One critic wrote that *Boléro* "departs from a thousand years of tradition."

I sat in my living room alone, listening. *Boléro* starts simply enough, a single flute accompanied by a snare drum: *da-da-da-dum, da-da-da-dum, dum-dum, da-da-da-dum.* The same musical clause repeats seventeen more times, each cycle adding instruments, growing louder and more insistent, until the entire orchestra roars in an overpowering finale of rhythm and sound. Musically, it was perfect for my ear. It had a structure that I could easily grasp and enough variation to hold my interest.

It took a lot to hold my interest; I was nearly deaf at the time. In 1964 my mother contracted rubella while pregnant with me. Hearing aids allowed me to understand speech well enough, but most music was lost on me. *Boléro* was one of the few pieces I actually enjoyed. A few years later, I bought the CD and played it so much it eventually grew pitted and scratched. It became my touchstone. Every time I tried out a new hearing aid, I'd check to see if *Boléro* sounded OK. If it didn't, the hearing aid went back.

And then, on July 7, 2001, at 10:30 A.M., I lost my ability to hear *Boléro*—and everything else. While I was waiting to pick up a rental car in Reno, I suddenly thought the battery in my hearing aid had died. I replaced it. No luck. I switched hearing aids. Nothing.

I got into my rental car and drove to the nearest emergency room. For reasons that are still unknown, my only functioning ear had suffered "sudden-onset deafness." I was reeling, trying to navigate in a world where the volume had been turned down to zero.

But there was a solution, a surgeon at Stanford Hospital told me a week later, speaking slowly so I could read his lips. I could have a computer surgically installed in my skull. A cochlear implant, as it is known, would trigger my auditory nerves with sixteen electrodes that snaked inside my inner ear. It seemed drastic, and the fifty-thousand-dollar price tag was a dozen times more expensive than a high-end hearing aid. I went home and cried. Then I said yes.

For the next two months, while awaiting surgery, I was totally deaf except for a thin trickle of sound from my right ear. I had long since become accustomed to not hearing my own voice when I spoke. It happened whenever I removed my hearing aid. But that sensation was as temporary as waking up without my glasses. Now, suddenly, the silence wasn't optional. At my job as a technical writer in Silicon Valley, I struggled at meetings. Using the phone was out of the question.

In early September, the surgeon drilled a tunnel through an inch and a half of bone behind my left ear and inserted the sixteen electrodes along the auditory nerve fibers in my cochlea. He hollowed a well in my skull about the size of three stacked quarters and snapped in the implant.

When the device was turned on a month after surgery, the first sentence I heard sounded like "Zzzzzz szz szvizzz ur brfzzzzzz?" My brain gradually learned how to interpret the alien signal. Before long, "Zzzzzz szz szvizzz ur brfzzzzzz?" became "What did you have for breakfast?" After months of practice, I could use the telephone again, even converse in loud bars and cafeterias. In many ways, my hearing was better than it had ever been. Except when I listened to music.

I could hear the drums of *Boléro* just fine. But the other instruments were flat and dull. The flutes and soprano saxophones sounded as though someone had clapped pillows over them. The oboes and violins had become groans. It was like walking color-blind through a Paul Klee exhibit. I played *Boléro* again and again, hoping that practice would bring it, too, back to life. It didn't.

The implant was embedded in my head; it wasn't some flawed hearing aid I could just send back. But it *was* a computer. Which meant that, at least in theory, its effectiveness was limited only by the ingenuity of software engineers. As researchers learn more about how the ear works, they continually revise cochlear implant software. Users await new releases with all the anticipation of Apple zealots lining up for the latest Mac OS.

About a year after I received the implant, I asked one implant engineer how much of the device's hardware capacity was being used. "Five percent, maybe." He shrugged. "Ten, tops."

I was determined to use that other 90 percent. I set out on a crusade to explore the edges of auditory science. For two years I tugged on the sleeves of scientists and engineers around the country, offering myself as a guinea pig for their experiments. I wanted to hear *Boléro* again.

HELEN KELLER FAMOUSLY SAID that if she had to choose between being deaf and being blind, she'd be blind, because while blindness cut her off from things, deafness cut her off from people. For centuries, the best available hearing aid was a horn, or ear trumpet, which people held to their ears to funnel in sound. In 1952 the first electronic hearing aid was developed. It worked by blasting amplified sound into a damaged ear. However, it (and the more advanced models that followed) could help only if the user had some residual hearing ability, just as glasses can help only those who still have some vision. Cochlear implants, on the other hand, bypass most of the ear's natural hearing mechanisms. The device's electrodes directly stimulate nerve endings in the ear, which transmit sound information to the brain. Since the surgery can eliminate any remaining hearing, implants are approved for use only in people who can't be helped by hearing aids. The first modern cochlear implants went on the market in Australia in 1982, and by 2004 approximately 82,500 people worldwide had been fitted with one.

When technicians activated my cochlear implant in October 2001, they gave me a pager-sized processor that decoded sound and sent it to a headpiece that clung magnetically to the implant underneath my skin. The headpiece contained a radio transmitter, which sent the processor's data to the implant at roughly 1 megabit per second. Sixteen electrodes curled up inside my cochlea strobed on and off to stimulate my auditory nerves. The processor's software gave me eight channels of auditory resolution, each representing a frequency range. The more channels the software delivers, the better the user can distinguish between sounds of different pitches.

Eight channels isn't much compared with the capacity of a normal ear, which has the equivalent of 3,500 channels. Still, eight works well enough for speech, which doesn't have much pitch variation. Music is another story. The lowest of my eight channels captured everything from 250 hertz (about middle C on the piano) to 494 hertz (close to the B above middle C), making it nearly impossible for me to distinguish among the eleven notes in that range. Every note that fell into a particular channel sounded the same to me.

So in mid-2002, nine months after activation, I upgraded to a program called Hi-Res, which gave me sixteen channels—double the resolution! An audiologist plugged my processor into her laptop and uploaded the new code. I suddenly had a better ear, without surgery. In theory, I would now be able to distinguish among tones five notes apart instead of eleven.

I eagerly plugged my Walkman into my processor and turned it on. *Boléro* did sound better. But after a day or two, I realized that "better" still wasn't good enough. The improvement was small, like being in that art gallery again and seeing only a gleam of pink here, a bit of blue there. I wasn't hearing the *Boléro* I remembered.

At a cochlear implant conference in 2003, I heard Jay Rubinstein, a surgeon and researcher at the University of Washington, say that it took at least one hundred channels of auditory information to make music pleasurable. My jaw dropped. No wonder. I wasn't even close.

A year later, I met Rubinstein at another conference, and he mentioned that there might be ways to bring music back to me. He told me about something called stochastic resonance; studies suggested that my music perception might be aided by deliberately adding noise to what I hear. He took a moment to give me a lesson in neural physiology. After a neuron fires, it goes dormant for a fraction of a second while it resets. During that phase, it misses any information that comes along. When an electrode zaps thousands of neurons at once, it forces them all to go dormant, making it impossible for them to receive pulses until they reset. That synchrony means I miss bits and pieces of information.

Desynchronizing the neurons, Rubinstein explained, would guarantee that they're never all dormant simultaneously. And the best way to get them out of sync is to beam random electrical noise at them. A few months later, Rubinstein arranged a demonstration.

An audiologist at the University of Iowa working with Rubenstein handed me a processor loaded with the stochastic-resonance software. The first thing I heard was a loud whoosh—the random noise. It sounded like a cranked-up electric fan. But in about thirty seconds, the noise went away. I was puzzled. "You've adapted to it," the technician told me. The nervous system can habituate to any kind of everyday sound, but it adjusts especially quickly to noise with no variation. Stochastic-resonance noise is so content-free that the brain tunes it out in seconds.

In theory, the noise would add just enough energy to incoming sound to make faint details audible. In practice, everything I heard became rough and gritty. My own voice sounded vibrato, mechanical, and husky—even a little querulous, as if I were perpetually whining.

We tried some quick tests to take my newly programmed ear out for a spin. It performed slightly better in some ways, slightly worse in others—but there was no dramatic improvement. The audiologist wasn't surprised. She told me that, in most cases, a test subject's brain will take weeks or even months to make sense of the additional information. Furthermore, the settings she chose were only an educated guess at what might work for my particular physiology. Everyone is different. Finding the right setting is like fishing for one particular cod in the Atlantic.

The university lent me the processor to test for a few months. As soon as I was back in the hotel, I tried my preferred version of *Boléro,* a 1982 recording conducted by Charles Dutoit with the Montréal Symphony Orchestra. It sounded different, but not better. Sitting at my keyboard, I sighed a little and tapped out an e-mail thanking Rubinstein and encouraging him to keep working on it.

Music depends on low frequencies for its richness and mellowness. The lowest-pitched string on a guitar vibrates at 83 hertz, but my Hi-Res software, like the eight-channel model, bottoms out at 250 hertz. I do hear something when I pluck a string, but it's not actually an 83-hertz sound. Even though the string is vibrating at eighty-three times per second, portions of it are vibrating faster, giving rise to higher-frequency notes called harmonics. The harmonics are what I hear.

The engineers haven't gone below 250 hertz because the world's low-pitched sounds—air conditioners, engine rumbles—interfere with speech perception. Furthermore, increasing the total frequency range means decreasing resolution, because each channel has to accommodate more frequencies. Since speech perception has been the main goal during decades of research, the engineers haven't given much thought to representing low frequencies. Until Philip Loizou came along.

Loizou and his team of postdocs at the University of Texas at Dallas are trying to figure out ways to give cochlear implant users access to more low frequencies. A week after my frustratingly inconclusive encounter with stochastic resonance, I traveled to Dallas and asked Loizou why the government would give him a grant to develop software that increases musical appreciation. "Music lifts up people's spirits, helps them forget things," he told me in his mild Greek accent. "The goal is to have the patient live a normal life, not to be deprived of anything."

Loizou is trying to negotiate a trade-off: narrowing low-frequency channels while widening higher-frequency channels. But his theories only hinted at what specific configurations might work best, so Loizou was systematically trying a range of settings to see which ones got the better results.

The team's software ran only on a desktop computer, so on my visit to Dallas I had to be plugged directly into the machine. After a round of testing, a postdoc assured me, they would run *Boléro* through their software and pipe it into my processor via Windows Media Player.

I spent two and a half days hooked up to the computer, listening to endless sequences of tones—none of it music—in a windowless cubicle. Which of two tones sounded lower? Which of two versions of "Twinkle, Twinkle, Little Star" was more recognizable? Did this string of notes sound like a march or a waltz? It was exacting, high-concentration work—like taking an eye exam that lasted for two days. My responses produced reams of data that they would spend hours analyzing.

Forty minutes before my cab back to the airport was due, we finished the last test and the postdoc fired up the programs he needed to play *Boléro.* Some of the lower pitches I'd heard in the previous two days had sounded rich and mellow, and I began thinking wistfully about those bassoons and oboes. I felt a rising sense of anticipation and hope.

I waited while the postdoc tinkered with the computer. And waited. Then I noticed the frustrated look of a man trying to get Windows to behave. "I do this all the time," he said, half to himself. Windows Media Player wouldn't play the file.

I suggested rebooting and sampling *Boléro* through a microphone. But the postdoc told me he couldn't do that in time for my plane. A later flight wasn't an option; I had to be back in the Bay Area. I was crushed. I walked out of the building with my shoulders slumped. Scientifically, the visit was a great success. But for me, it was a failure. On the flight home, I plugged myself into my laptop and listened sadly to *Boléro* with Hi-Res. It was like eating cardboard.

IT'S JUNE 2005, a few weeks after my visit to Dallas, and I'm ready to try again. A team of engineers at Advanced Bionics, one of three companies in the world that makes bionic ears, is working on a new software algorithm for so-called virtual channels. I hop on a flight to their Los Angeles headquarters, my CD player in hand.

My implant has sixteen electrodes, but the virtual-channels software will make my hardware act like there are actually 121. Manipulating the flow of electricity to target neurons between each electrode creates the illusion of seven new electrodes between each actual pair, similar to the way an audio engineer can make a sound appear to emanate from between two speakers. Jay Rubinstein had told me two years ago that it would take at least one hundred channels to create good music perception. I'm about to find out if he's right.

I'm sitting across a desk from Gulam Emadi, an Advanced Bionics researcher. He and an audiologist are about to fit me with the new software. Leo Litvak, who has spent three years developing the program, comes in to say hello. He's one of those people of whom others often say, "If Leo can't do it, it probably can't be done." And yet it would be hard to find a more modest person. Were it not for his clothes, which mark him as an Orthodox Jew, he would simply disappear in a roomful of people. Litvak tilts his head and smiles hello, shyly glances at Emadi's laptop, and sidles out.

At this point, I'm rationing my emotions like Spock. Hi-Res was a disappointment. Stochastic resonance remains a big if. The low-frequency experiment in Dallas was a bust. Emadi dinks with his computer and hands me my processor with the new software in it. I plug it into myself, plug my CD player into it, and press Play.

Boléro starts off softly and slowly, meandering like a breeze through the trees. *Da-da-da-dum, da-da-da-dum, dum-dum, da-da-da-dum.* I close my eyes to focus, switching between Hi-Res and the new software every twenty or thirty seconds by thumbing a blue dial on my processor.

My God, the oboes d'amore do sound richer and warmer. I let out a long, slow breath, coasting down a river of sound, waiting for the soprano saxophones and the piccolos. They'll come in around six minutes into the piece—and it's only then that I'll know if I've truly got it back.

As it turns out, I couldn't have chosen a better piece of music for testing new implant software. Some biographers have suggested that *Boléro*'s obsessive repetition is rooted in the neurological problems Ravel had started to exhibit in 1927, a year before he composed the piece. It's still up for debate whether he had early-onset Alzheimer's, a left-hemisphere brain lesion, or something else.

But *Boléro*'s obsessiveness, whatever its cause, is just right for my deafness. Over and over the theme repeats, allowing me to listen for specific details in each cycle.

At 5:59, the soprano saxophones leap out bright and clear, arcing above the snare drum. I hold my breath.

At 6:39, I hear the piccolos. For me, the stretch between 6:39 and 7:22 is the most *Boléro* of *Boléro*, the part I wait for each time. I concentrate. It sounds . . . *right.*

Hold on. Don't jump to conclusions. I backtrack to 5:59 and switch to Hi-Res. That heart-stopping leap has become an asthmatic whine. I backtrack again and switch to the new software. And there it is again, that exultant ascent. I can hear *Boléro*'s force, its intensity and passion. My chin starts to tremble.

I open my eyes, blinking back tears. "Congratulations," I say to Emadi. "You have done it." And I reach across the desk with absurd formality and shake his hand.

There's more technical work to do, more progress to be made, but I'm completely shattered. I keep zoning out and asking Emadi to repeat things. He passes me a box of tissues. I'm overtaken by a vast sensation of surprise. I did it. For years I pestered researchers and asked questions. Now I'm running 121 channels and I can hear music again.

That evening, in the airport, sitting numbly at the gate, I listen to *Boléro* again. I'd never made it through more than three or four minutes of the piece on Hi-Res before getting bored and turning it off. Now, I listen to the end, following the narrative, hearing again its holy madness.

I pull out the Advanced Bionics T-shirt that the team gave me and dab at my eyes.

DURING THE NEXT FEW DAYS I walk around in a haze of disbelief, listening to *Boléro* over and over to prove to myself that I really am hearing it again. But *Boléro* is just one piece of music. Jonathan Berger, head of Stanford's music department, tells me in an e-mail,

"There's not much of interest in terms of structure—it's a continuous crescendo, no surprises, no subtle interplay between development and contrast."

"In fact," he continues, "Ravel was not particularly happy that this study in orchestration became his big hit. It pales in comparison to any of his other music in terms of sophistication, innovation, grace, and depth."

So now it's time to try out music with sophistication, innovation, grace, and depth. But I don't know where to begin. I need an expert with first-rate equipment, a huge music collection, and the ability to pick just the right pieces for my newly reprogrammed ear. I put the question to craigslist—"Looking for a music geek." Within hours, I hear from Tom Rettig, a San Francisco music producer.

In his studio, Rettig plays me Ravel's *String Quartet in F Major* and Philip Glass's *String Quartet No. 5*. I listen carefully, switching between the old software and the new. Both compositions sound enormously better on 121 channels. But when Rettig plays music with vocals, I discover that having 121 channels hasn't solved all my problems. While the crescendos in Dulce Pontes's *Canção do Mar* sound louder and clearer, I hear only white noise when her voice comes in. Rettig figures that relatively simple instrumentals are my best bet—pieces where the instruments don't overlap too much—and that flutes and clarinets work well for me. Cavalcades of brass tend to overwhelm me and confuse my ear.

And some music just leaves me cold: I can't even get through Kraftwerk's *Tour de France*. I wave impatiently to Rettig to move on. (Later, a friend tells me it's not the software—Kraftwerk is just dull. It makes me think that for the first time in my life I might be developing a taste in music.)

Listening to *Boléro* more carefully in Rettig's studio reveals other bugs. The drums sound squeaky—how can drums squeak?—and in the frenetic second half of the piece, I still have trouble separating the instruments.

After I get over the initial awe of hearing music again, I discover that it's harder for me to understand ordinary speech than it was before I went to virtual channels. I report this to Advanced Bionics, and my complaint is met by a rueful shaking of heads. I'm not the first person to say that, they tell me. The idea of virtual channels is a breakthrough, but the technology is still in the early stages of development.

But I no longer doubt that incredible things can be done with that unused 90 percent of my implant's hardware capacity. Tests conducted a month after my visit to Advanced Bionics show that my ability to discriminate among notes has improved considerably. With Hi-Res, I was able to identify notes only when they were at least 70 hertz apart. Now, I can hear notes that are only 30 hertz apart. It's like going from being able to tell the difference between red and blue to being able to distinguish between aquamarine and cobalt.

My hearing is no longer limited by the physical circumstances of my body. While my friends' ears will inevitably decline with age, mine will only get better.

ALAN WEISMAN

Earth Without People

FROM *DISCOVER*

How would our urban landscapes be transformed if all human life were suddenly to vanish? After thousands of years, bridges would collapse, wildlife would roam streets overgrown with vegetation, and the seemingly invincible cockroach would only thrive in warm climates. Alan Weisman gives a fascinating preview of this depopulated world.

Given the mounting toll of fouled oceans, overheated air, missing topsoil, and mass extinctions, we might sometimes wonder what our planet would be like if humans suddenly disappeared. Would Superfund sites revert to Gardens of Eden? Would the seas again fill with fish? Would our concrete cities crumble to dust from the force of tree roots, water, and weeds? How long would it take for our traces to vanish? And if we could answer such questions, would we be more in awe of the changes we have wrought or of nature's resilience?

A good place to start searching for answers is in Korea, in the 155-mile-long, 2.5-mile-wide mountainous Demilitarized Zone, or DMZ, set up by the armistice ending the Korean War. Aside from rare military patrols or desperate souls fleeing North Korea, humans have barely set foot in the strip since 1953. Before that, for five thousand years, the area was populated by rice farmers who carved the land into paddies. Today those paddies have become barely discernible, transformed into pockets of marsh, and the new occupants of these lands arrive as dazzling white squadrons of red-crowned cranes that glide over the bulrushes in perfect formation, touching down so lightly that they detonate no land mines. Next to whooping cranes, they are the rarest such birds on Earth. They winter in the DMZ alongside the endangered white-naped cranes, revered in Asia as sacred portents of peace.

If peace is ever declared, suburban Seoul, which has rolled ever northward in recent decades, is poised to invade such tantalizing real estate. On the other side, the North Koreans are building an industrial megapark. This has spurred an international coalition of scientists called the DMZ Forum to try to consecrate the area for a peace park and nature preserve. Imagine it as "a Korean Gettysburg and Yosemite rolled together," says Harvard University biologist Edward O. Wilson, who believes that tourism revenues could trump those from agriculture or development.

As serenely natural as the DMZ now is, it would be far different if people throughout Korea suddenly disappeared. The habitat would not revert to a truly natural state until the dams that now divert rivers to slake the needs of Seoul's more than twenty million inhabitants failed—a century or two after the humans had gone. But in the meantime, says Wilson, many creatures would flourish. Otters, Asiatic black bears, musk deer, and the nearly vanquished Amur leopard would spread into slopes reforested with young daimyo oak and bird cherry. The few Siberian tigers that still prowl the North Korean–Chinese borderlands would multiply and fan across Asia's temperate zones. "The wild carnivores would make short work of livestock," he says. "Few domestic animals would remain after a couple of hundred years. Dogs would go feral, but they wouldn't last long: They'd never be able to compete."

If people were no longer present anywhere on Earth, a worldwide shakeout would follow. From zebra mussels to fire ants to crops to kudzu, exotics would battle with natives. In time, says Wilson, all human attempts to improve on nature, such as our painstakingly bred horses, would revert to their origins. If horses survived at all, they would devolve back to Przewalski's horse, the only true wild horse, still found in the Mongolian steppes. "The plants, crops, and animal species man has wrought by his own hand would be wiped out in a century or two," Wilson says. In a few thousand years, "the world would mostly look as it did before humanity came along—like a wilderness."

THE NEW WILDERNESS WOULD CONSUME CITIES, much as the jungle of northern Guatemala consumed the Mayan pyramids and megalopolises of overlapping city-states. From A.D. 800 to 900, a combination of drought and internecine warfare over dwindling farmland brought two thousand years of civilization crashing down. Within ten centuries, the jungle swallowed all.

Mayan communities alternated urban living with fields sheltered by forests, in contrast with today's paved cities, which are more like man-made deserts. However, it wouldn't take long for nature to undo even the likes of a New York City. Jameel Ahmad, civil engineering department chair at Cooper Union College in New York City, says repeated freezing and thawing common in months like March and November would split cement within a decade, allowing water to seep in. As it, too, froze and expanded, cracks would widen. Soon, weeds such as mustard and goosegrass would invade. With nobody to trample seedlings, New York's prolific exotic, the Chinese ailanthus tree, would take over. Within five years, says Dennis Stevenson, senior curator at the New York Botanical Garden, ailanthus roots would heave up sidewalks and split sewers.

That would exacerbate a problem that already plagues New York—rising groundwater. There's little soil to absorb it or vegetation to transpire it, and buildings block the sunlight that could evaporate it. With the power off, pumps that keep subways from flooding would be stilled. As water sluiced away soil beneath pavement, streets would crater.

Eric Sanderson of the Bronx Zoo Wildlife Conservation Society heads the Mannahatta Project, a virtual re-creation of pre-1609 Manhattan. He says there were thirty to forty streams in Manhattan when the Dutch first arrived. If New Yorkers disappeared, sewers would clog, some natural watercourses would reappear, and others would form. Within twenty years, the water-soaked steel columns that support the street above the East Side's subway tunnels would corrode and buckle, turning Lexington Avenue into a river.

New York's architecture isn't as flammable as San Francisco's clapboard Victorians, but within two hundred years, says Steven Clemants, vice president of the Brooklyn Botanic Garden, tons of leaf litter would overflow gutters as pioneer weeds gave way to colonizing native oaks and maples in city parks. A dry lightning strike, igniting decades of uncut, knee-high Central Park grass, would spread flames through town.

As lightning rods rusted away, roof fires would leap among buildings into paneled offices filled with paper. Meanwhile, native Virginia creeper and poison ivy would claw at walls covered with lichens, which thrive in the absence of air pollution. Wherever foundations failed and buildings tumbled, lime from crushed concrete would raise soil pH, inviting buckthorn and birch. Black locust and autumn olive trees would fix nitrogen, allowing more goldenrods, sunflowers, and white snakeroot to move in along with apple trees, their seeds expelled by proliferating birds. Sweet carrots would quickly devolve to their wild form, unpalatable Queen Anne's lace; white broccoli; cabbage; brussels sprouts; and cauliflower would regress to the same unrecognizable broccoli ancestor.

Unless an earthquake strikes New York first, bridges spared yearly applications of road salt would last a few hundred years before their stays and bolts gave way (last to fall would be Hell Gate Arch, built for railroads and easily good for another thousand years). Coyotes would invade Central Park, and deer, bears, and finally wolves would follow. Ruins would echo the love song of frogs breeding in streams stocked with alewives, herring, and mussels dropped by seagulls. Missing, however, would be all fauna that have adapted to humans. The invincible cockroach, an insect that originated in the hot climes of Africa, would succumb in unheated buildings. Without garbage, rats would starve or serve as lunch for peregrine falcons and red-tailed hawks. Pigeons would genetically revert back to the rock doves from which they sprang.

It's unclear how long animals would suffer from the urban legacy of concentrated heavy metals. Over many centuries, plants would take these up, recycle, redeposit, and gradually dilute them. The time bombs left in petroleum tanks, chemical plants, power plants, and dry-cleaning plants might poison the earth beneath them for eons. One intriguing example is the former Rocky Mountain Arsenal next to Denver International Airport. There a chemical weapons plant produced mustard and nerve gas, incendiary bombs, napalm, and after World War II, pesticides. In 1984 it was considered by the arsenal commander to be the most contaminated spot in the United States. Today it is a national wildlife refuge, home to bald eagles that feast on its prodigious prairie dog population.

However, it took more than $130 million and a lot of man-hours to drain and seal the arsenal's lake, in which ducks once died minutes after landing and the aluminum bottoms of boats sent to fetch their carcasses rotted within a month. In a world with no one left to bury the bad stuff, decaying chemical containers would slowly expose their lethal contents. Places like the Indian Point nuclear power plant, thirty-five miles north of Times Square, would dump radioactivity into the Hudson long after the lights went out.

Old stone buildings in Manhattan, such as Grand Central Station or the Metropolitan Museum of Art, would outlast every modern glass box, especially with no more acid rain to pock their marble. Still, at some point thousands of years hence, the last stone walls—perhaps chunks of St. Paul's Chapel on Wall Street, built in 1766 from Manhattan's own hard schist—would fall. Three times in the past one hundred thousand years, glaciers have scraped New York clean, and they'll do so again. The mature hardwood forest would be mowed down. On Staten Island, Fresh Kills's four giant mounds of trash would be flattened, their vast accumulation of stubborn PVC plastic and glass ground to powder. After the ice receded, an unnatural concentration of reddish metal—remnants of wiring and plumbing—would remain buried in layers. The next toolmaker to arrive or evolve might discover it and use it, but there would be nothing to indicate who had put it there.

BEFORE HUMANS APPEARED, an oriole could fly from the Mississippi to the Atlantic and never alight on anything other than a treetop. Unbroken forest blanketed Europe from the Urals to the English Channel. The last remaining fragment of that primeval European wilderness—half a million acres of woods straddling the border between Poland and Belarus, called the Bialowieza Forest—provides another glimpse of how the world would look if we were gone. There, relic groves of huge ash and linden trees rise 138 feet above an understory of hornbeams, ferns, swamp alders, massive birches, and crockery-size fungi. Norway spruces, shaggy as Methuselah, stand even taller. Five-century-old oaks grow so immense that great spotted woodpeckers stuff whole spruce cones in their three-inch-deep bark furrows. The woods carry pygmy owl whistles, nutcracker croaks, and wolf howls. Fragrance wafts from eons of mulch.

High privilege accounts for such unbroken antiquity. During the fourteenth century, a Lithuanian duke declared it a royal hunting preserve. For centuries it stayed that way. Eventually, the forest was subsumed by Russia and in 1888 became the private domain of the czars. Occupying Germans took lumber and slaughtered game during World War I, but a pristine core was left intact, which in 1921 became a Polish national park. Timber pillaging resumed briefly under the

Soviets, but when the Nazis invaded, nature fanatic Hermann Göring decreed the entire preserve off limits. Then, following World War II, a reportedly drunken Joseph Stalin agreed one evening in Warsaw to let Poland retain two-fifths of the forest.

To realize that all of Europe once looked like this is startling. Most unexpected of all is the sight of native bison. Just six hundred remain in the wild, on both sides of an impassable iron curtain erected by the Soviets in 1980 along the border to thwart escapees to Poland's renegade Solidarity movement. Although wolves dig under it, and roe deer are believed to leap over it, the herd of the largest of Europe's mammals remains divided, and thus its gene pool. Belarus, which has not removed its statues of Lenin, has no specific plans to dismantle the fence. Unless it does, the bison may suffer genetic degradation, leaving them vulnerable to a disease that would wipe them out.

If the bison herd withers, they would join all the other extinct megafauna that even our total disappearance could never bring back. In a glass case in his laboratory, paleoecologist Paul S. Martin at the University of Arizona keeps a lump of dried dung he found in a Grand Canyon cave, left by a sloth weighing two hundred pounds. That would have made it the smallest of several North American ground sloth species present when humans first appeared on this continent. The largest was as big as an elephant and lumbered around by the thousands in the woodlands and deserts of today's United States. What we call pristine today, Martin says, is a poor reflection of what would be here if *Homo sapiens* had never evolved.

"America would have three times as many species of animals over one thousand pounds as Africa does today," he says. An amazing megafaunal menagerie roamed the region: giant armadillos resembling armor-plated autos; bears twice the size of grizzlies; the hoofed, herbivorous toxodon, big as a rhinoceros; and saber-toothed tigers. A dozen species of horses were here, as well as the camel-like litoptern, giant beavers, giant peccaries, woolly rhinos, mammoths, and mastodons. Climate change and imported disease may have killed them, but most paleontologists accept the theory Martin advocates:

"When people got out of Africa and Asia and reached other parts of the world, all hell broke loose." He is convinced that people were responsible for the mass extinctions because they commenced with human arrival everywhere: first, in Australia sixty thousand years ago, then mainland America thirteen thousand years ago, followed by the Caribbean islands six thousand years ago, and Madagascar two thousand years ago.

Yet one place on Earth did manage to elude the intercontinental holocaust: the oceans. Dolphins and whales escaped for the simple reason that prehistoric people could not hunt enough giant marine mammals to have a major impact on the population. "At least a dozen species in the ocean Columbus sailed were bigger than his biggest ship," says marine paleocologist Jeremy Jackson of the Smithsonian Tropical Research Institute in Panama. "Not only mammals—the sea off Cuba was so thick with one-thousand-pound green turtles that his boats practically ran aground on them." This was a world where ships collided with schools of whales and where sharks were so abundant they would swim up rivers to prey on cattle. Reefs swarmed with eight-hundred-pound goliath grouper, not just today's puny aquarium species. Cod could be fished from the sea in baskets. Oysters filtered all the water in Chesapeake Bay every five days. The planet's shores teemed with millions of manatees, seals, and walrus.

Within the past century, however, humans have flattened the coral reefs on the continental shelves and scraped the sea grass beds bare; a dead zone bigger than New Jersey grows at the mouth of the Mississippi; all the world's cod fisheries have collapsed. What Pleistocene humans did in fifteen hundred years to terrestrial life, modern man has done in mere decades to the oceans—"almost," Jackson says. Despite mechanized overharvesting, satellite fish tracking, and prolonged butchery of sea mammals, the ocean is still bigger than we are. "It's not like the land," he says. "The great majority of sea species are badly depleted, but they still exist. If people actually went away, most could recover."

Even if global warming or ultraviolet radiation bleaches the Great

Barrier Reef to death, Jackson says, "it's only seven thousand years old. New reefs have had to form before. It's not like the world is a constant place." Without people, most excess industrial carbon dioxide would dissipate within two hundred years, cooling the atmosphere. With no further chlorine and bromine leaking skyward, within decades the ozone layer would replenish, and ultraviolet damage would subside. Eventually, heavy metals and toxins would flush through the system; a few intractable PCBs might take a millennium.

During that same span, every dam on Earth would silt up and spill over. Rivers would again carry nutrients seaward, where most life would be, as it was long before vertebrates crawled onto the shore. Eventually, that would happen again. The world would start over.

ELIZABETH KOLBERT

The Curse of Akkad

FROM *THE NEW YORKER*

In early 2006, James Hansen, a NASA environmental scientist, announced that NASA was suppressing his comments about climate change. This embarrassing incident could not quiet the disturbing message of Hansen's research: that global warming could cause world-wide catastrophe in this century. Using Hansen to guide her through the interpretation of these unsettling trends as well as the fall of ancient civilizations, Elizabeth Kolbert explains why the alarm has been sounded.

The world's first empire was established forty-three hundred years ago, between the Tigris and Euphrates Rivers. The details of its founding, by Sargon of Akkad, have come down to us in a form somewhere between history and myth. Sargon—Sharru-kin, in the language of Akkadian—means "true king"; almost certainly, though, he was a usurper. As a baby, Sargon was said to have been discovered, Moses-like, floating in a basket. Later, he became cupbearer to the ruler of Kish, one of ancient Babylonia's most powerful cities. Sargon dreamed that his master, Ur-Zababa, was about to be drowned by the goddess Inanna in a river of blood. Hearing about the dream, Ur-Zababa decided to have Sargon eliminated. How this plan failed is unknown; no text relating the end of the story has ever been found.

Until Sargon's reign, Babylonian cities like Kish, and also Ur and Uruk and Umma, functioned as independent city-states. Sometimes they formed brief alliances—cuneiform tablets attest to strategic marriages celebrated and diplomatic gifts exchanged—but mostly they seem to have been at war with one another. Sargon first subdued Babylonia's fractious cities, then went on to conquer, or at least sack, lands like Elam, in present-day Iran. He presided over his empire from the city of Akkad, the ruins of which are believed to lie south of Baghdad. It was written that "daily five thousand four hundred men ate at his presence," meaning, presumably, that he maintained a huge standing army. Eventually, Akkadian hegemony extended as far as the Khabur plains, in northeastern Syria, an area prized for its grain production. Sargon came to be known as "king of the world"; later, one of his descendants enlarged this title to "king of the four corners of the universe."

Akkadian rule was highly centralized, and in this way anticipated the administrative logic of empires to come. The Akkadians levied taxes, then used the proceeds to support a vast network of local bureaucrats. They introduced standardized weights and measures—the *gur* equaled roughly three hundred litres—and imposed a uniform dating system, under which each year was assigned the name of a major event that had recently occurred: for instance, "the year that Sargon destroyed the city of Mari." Such was the level of systematization that even the shape and the layout of accounting tablets were imperially prescribed. Akkad's wealth was reflected in, among other things, its art work, the refinement and naturalism of which were unprecedented.

Sargon ruled, supposedly, for fifty-six years. He was succeeded by his two sons, who reigned for a total of twenty-four years, and then by a grandson, Naramsin, who declared himself a god. Naramsin was, in turn, succeeded by his son. Then, suddenly, Akkad collapsed. During one three-year period, four men each, briefly, claimed the throne. "Who was king? Who was not king?" the register known as the Sumerian King List asks, in what may be the first recorded instance of political irony.

The lamentation "The Curse of Akkad" was written within a century of the empire's fall. It attributes Akkad's demise to an outrage against the gods. Angered by a pair of inauspicious oracles, Naram-sin plunders the temple of Enlil, the god of wind and storms, who, in retaliation, decides to destroy both him and his people:

> For the first time since cities were built and founded,
> The great agricultural tracts produced no grain,
> The inundated tracts produced no fish,
> The irrigated orchards produced neither syrup nor wine,
> The gathered clouds did not rain, the *masgurum* did not grow.
> At that time, one shekel's worth of oil was only one-half quart,
> One shekel's worth of grain was only one-half quart. . . .
> These sold at such prices in the markets of all the cities!
> He who slept on the roof, died on the roof,
> He who slept in the house, had no burial,
> People were flailing at themselves from hunger.

For many years, the events described in "The Curse of Akkad" were thought, like the details of Sargon's birth, to be purely fictional.

IN 1978, after scanning a set of maps at Yale's Sterling Memorial Library, a university archaeologist named Harvey Weiss spotted a promising-looking mound at the confluence of two dry riverbeds in the Khabur plains, near the Iraqi border. He approached the Syrian government for permission to excavate the mound, and, somewhat to his surprise, it was almost immediately granted. Soon, he had uncovered a lost city, which in ancient times was known as Shekhna and today is called Tell Leilan.

Over the next ten years, Weiss, working with a team of students and local laborers, proceeded to uncover an acropolis, a crowded residential neighborhood reached by a paved road, and a large block of grain-storage rooms. He found that the residents of Tell Leilan had raised barley and several varieties of wheat, that they had used carts to transport their crops, and that in their writing they had imitated the style of their more sophisticated neighbors to the south. Like most cities in the region at the time, Tell Leilan had a rigidly organized, state-run economy: people received rations—so many litres of barley and so many of oil—based on how old they were and what kind of work they performed. From the time of the Akkadian empire, thousands of similar potsherds were discovered, indicating that residents had received their rations in mass-produced, one-litre vessels. After examining these and other artifacts, Weiss constructed a time line of the city's history, from its origins as a small farming village (around 5000 BC), to its growth into an independent city of some thirty thousand people (2600 BC), and on to its reorganization under imperial rule (2300 BC).

Wherever Weiss and his team dug, they also encountered a layer of dirt that contained no signs of human habitation. This layer, which was more than three feet deep, corresponded to the years 2200 to 1900 BC, and it indicated that, around the time of Akkad's fall, Tell Leilan had been completely abandoned. In 1991 Weiss sent soil samples from Tell Leilan to a lab for analysis. The results showed that, around the year 2200 BC, even the city's earthworms had died out. Eventually, Weiss came to believe that the lifeless soil of Tell Leilan and the end of the Akkadian empire were products of the same phenomenon—a drought so prolonged and so severe that, in his words, it represented an example of "climate change."

Weiss first published his theory, in the journal *Science* in August 1993. Since then, the list of cultures whose demise has been linked to climate change has continued to grow. They include the Classic Mayan civilization, which collapsed at the height of its development, around AD 800; the Tiwanaku civilization, which thrived near Lake

Titicaca, in the Andes, for more than a millennium, then disintegrated around AD 1100; and the Old Kingdom of Egypt, which collapsed around the same time as the Akkadian empire. (In an account eerily reminiscent of "The Curse of Akkad," the Egyptian sage Ipuwer described the anguish of the period: "Lo, the desert claims the land. Towns are ravaged. . . . Food is lacking. . . . Ladies suffer like maidservants. Lo, those who were entombed are cast on high grounds.") In each of these cases, what began as a provocative hypothesis has, as new information has emerged, come to seem more and more compelling. For example, the notion that Mayan civilization had been undermined by climate change was first proposed in the late 1980s, at which point there was little climatological evidence to support it. Then, in the mid-1990s, American scientists studying sediment cores from Lake Chichancanab, in north-central Yucatán, reported that precipitation patterns in the region had indeed shifted during the ninth and tenth centuries, and that this shift had led to periods of prolonged drought. More recently, a group of researchers examining ocean-sediment cores collected off the coast of Venezuela produced an even more detailed record of rainfall in the area. They found that the region experienced a series of severe, "multi-year drought events" beginning around AD 750. The collapse of the Classic Mayan civilization, which has been described as "a demographic disaster as profound as any other in human history," is thought to have cost millions of lives.

The climate shifts that affected past cultures predate industrialization by hundreds—or, in the case of the Akkadians, thousands—of years. They reflect the climate system's innate variability and were caused by forces that, at this point, can only be guessed at. By contrast, the climate shifts predicted for the coming century are attributable to forces that are now well known. Exactly how big these shifts will be is a matter of both intense scientific interest and the greatest possible historical significance. In this context, the discovery that large and sophisticated cultures have already been undone by climate change presents what can only be called an uncomfortable precedent.

THE GODDARD INSTITUTE FOR SPACE STUDIES, or GISS, is situated just south of Columbia University's main campus, at the corner of Broadway and West 112th Street. The institute is not well marked, but most New Yorkers would probably recognize the building: its ground floor is home to Tom's Restaurant, the coffee shop made famous by *Seinfeld*.

GISS, an outpost of NASA, started out, forty-four years ago, as a planetary-research center; today, its major function is making forecasts about climate change. GISS employs about a hundred and fifty people, many of whom spend their days working on calculations that may—or may not—end up being incorporated in the institute's climate model. Some work on algorithms that describe the behavior of the atmosphere, some on the behavior of the oceans, some on vegetation, some on clouds, and some on making sure that all these algorithms, when they are combined, produce results that seem consistent with the real world. (Once, when some refinements were made to the model, rain nearly stopped falling over the rain forest.) The latest version of the GISS model, called ModelE, consists of a hundred and twenty-five thousand lines of computer code.

GISS's director, James Hansen, occupies a spacious, almost comically cluttered office on the institute's seventh floor. (I must have expressed some uneasiness the first time I visited him, because the following day I received an e-mail assuring me that the office was "a lot better organized than it used to be.") Hansen, who is sixty-three, is a spare man with a lean face and a fringe of brown hair. Although he has probably done as much to publicize the dangers of global warming as any other scientist, in person he is reticent almost to the point of shyness. When I asked him how he had come to play such a prominent role, he just shrugged. "Circumstances," he said.

Hansen first became interested in climate change in the mid-1970s. Under the direction of James Van Allen (for whom the Van Allen radiation belts are named), he had written his doctoral dissertation on the climate of Venus. In it, he had proposed that the planet, which has an average surface temperature of 867 degrees Fahrenheit, was kept warm by a smoggy haze; soon afterward, a space probe showed that Venus was actually insulated by an atmosphere that consists of 96 percent carbon dioxide. When solid data began to show what was happening to greenhouse-gas levels on Earth, Hansen became, in his words, "captivated." He decided that a planet whose atmosphere could change in the course of a human lifetime was more interesting than one that was going to continue, for all intents and purposes, to broil away forever. A group of scientists at NASA had put together a computer program to try to improve weather forecasting using satellite data. Hansen and a team of half a dozen other researchers set out to modify it, in order to make longer-range forecasts about what would happen to global temperatures as greenhouse gases continued to accumulate. The project, which resulted in the first version of the GISS climate model, took nearly seven years to complete.

At that time, there was little empirical evidence to support the notion that the earth was warming. Instrumental temperature records go back, in a consistent fashion, only to the mid-nineteenth century. They show that average global temperatures rose through the first half of the twentieth century, then dipped in the 1950s and 1960s. Nevertheless, by the early 1980s Hansen had gained enough confidence in his model to begin to make a series of increasingly audacious predictions. In 1981, he forecast that "carbon dioxide warming should emerge from the noise of natural climate variability" around the year 2000. During the exceptionally hot summer of 1988, he appeared before a Senate subcommittee and announced that he was "99 percent" sure that "global warming is affecting our planet now."

And in the summer of 1990 he offered to bet a roomful of fellow-scientists a hundred dollars that either that year or one of the following two years would be the warmest on record. To qualify, the year would have to set a record not only for land temperatures but also for sea-surface temperatures and for temperatures in the lower atmosphere. Hansen won the bet in six months.

Like all climate models, GISS's divides the world into a series of boxes. Thirty-three hundred and twelve boxes cover the earth's surface, and this pattern is repeated twenty times moving up through the atmosphere, so that the whole arrangement might be thought of as a set of enormous checkerboards stacked on top of one another. Each box represents an area of four degrees latitude by five degrees longitude. (The height of the box varies depending on altitude.) In the real world, of course, such a large area would have an incalculable number of features; in the world of the model, features such as lakes and forests and, indeed, whole mountain ranges are reduced to a limited set of properties, which are then expressed as numerical approximations. Time in this grid world moves ahead for the most part in discrete, half-hour intervals, meaning that a new set of calculations is performed for each box for every thirty minutes that is supposed to have elapsed in actuality. Depending on what part of the globe a box represents, these calculations may involve dozens of different algorithms, so that a model run that is supposed to simulate climate conditions over the next hundred years involves more than a quadrillion separate operations. A single run of the GISS model, done on a supercomputer, usually takes about a month.

Very broadly speaking, there are two types of equations that go into a climate model. The first group expresses fundamental physical principles, like the conservation of energy and the law of gravity. The second group describes—the term of art is "parameterize"—patterns and interactions that have been observed in nature but may be only partly understood, or processes that occur on a small scale, and have to be averaged out over huge spaces. Here, for example, is a tiny piece of ModelE, written in the computer language FORTRAN, which deals with the formation of clouds:

```
C**** COMPUTE THE AUTOCONVERSION
RATE OF CLOUD WATER TO PRECIPITATION
    RHO=1.E5*PL(L)/(RGAS*TL(L))
```

```
TEM=RHO*WMX(L)/(WCONST*FCLD+
    1.E-20)
IF(LHX.EQ.LHS) TEM=RHO*WMX(L)/
    (WMUI*FCLD+1.E-20)
TEM=TEM*TEM
IF(TEM.GT.10.) TEM=10.
CM1=CM0
IF(BANDF) CM1=CM0*CBF
IF(LHX.EQ.LHS) CM1=CM0
CM=CM1*(1.-1./EXP(TEM*TEM))+1.
    *100.*(PREBAR(L+1)+
*PRECNVL(L+1)*BYDTSRC)
IF(CM.GT.BYDTSRC) CM=BYDTSRC
PREP(L)=WMX(L)*CM
END IF
C**** FORM CLOUDS ONLY IF RH GT RH00
  219 IF(RH1(L).LT.RH00(L)) GO TO 220.
```

All climate models treat the laws of physics in the same way, but since they parameterize phenomena like cloud formation differently, they come up with different results. (At this point, there are some fifteen major climate models in operation around the globe.) Also, because the real-world forces influencing the climate are so numerous, different models tend, like medical students, to specialize in different processes. GISS's model, for example, specializes in the behavior of the atmosphere, other models in the behavior of the oceans, and still others in the behavior of land surfaces and ice sheets.

Last fall, I attended a meeting at GISS which brought together members of the institute's modeling team. When I arrived, about twenty men and five women were sitting in battered chairs in a conference room across from Hansen's office. At that particular moment, the institute was performing a series of runs for the UN Intergovernmental Panel on Climate Change. The runs were overdue, and apparently the IPCC was getting impatient. Hansen flashed a series of charts on a screen on the wall summarizing some of the results obtained so far.

The obvious difficulty in verifying any particular climate model or climate-model run is the prospective nature of the results. For this reason, models are often run into the past, to see how well they reproduce trends that have already been observed. Hansen told the group that he was pleased with how ModelE had reproduced the aftermath of the eruption of Mt. Pinatubo, in the Philippines, which took place in June of 1991. Volcanic eruptions release huge quantities of sulfur dioxide—Pinatubo produced some twenty million tons of the gas—which, once in the stratosphere, condenses into tiny sulfate droplets. These droplets, or aerosols, tend to cool the earth by reflecting sunlight back into space. (Man-made aerosols, produced by burning coal, oil, and biomass, also reflect sunlight and are a countervailing force to greenhouse warming, albeit one with serious health consequences of its own.) This cooling effect lasts as long as the aerosols remain suspended in the atmosphere. In 1992, global temperatures, which had been rising sharply, fell by half of a degree. Then they began to climb again. ModelE had succeeded in simulating this effect to within nine-hundredths of a degree. "That's a pretty nice test," Hansen observed laconically.

ONE DAY, when I was talking to Hansen in his office, he pulled a pair of photographs out of his briefcase. The first showed a chubby-faced five-year-old girl holding some miniature Christmas-tree lights in front of an even chubbier-faced five-month-old baby. The girl, Hansen told me, was his granddaughter Sophie and the boy was his new grandson, Connor. The caption on the first picture read, "Sophie explains greenhouse warming." The caption on the second photograph, which showed the baby smiling gleefully, read, "Connor gets it."

When modelers talk about what drives the climate, they focus on what they call "forcings." A forcing is any ongoing process or discrete event that alters the energy of the system. Examples of natural forcings include, in addition to volcanic eruptions, periodic shifts in the earth's orbit and changes in the sun's output, like those linked to sunspots. Many climate shifts of the past have no known forcing associated with them; for instance, no one is certain what brought about the so-called Little Ice Age, which began in Europe some five hundred years ago. A very large forcing, meanwhile, should produce a commensurately large—and obvious—effect. One GISS scientist put it to me this way: "If the sun went supernova, there's no question that we could model what would happen."

Adding carbon dioxide, or any other greenhouse gas, to the atmosphere by, say, burning fossil fuels or leveling forests is, in the language of climate science, an anthropogenic forcing. Since preindustrial times, the concentration of CO_2 in the earth's atmosphere has risen by roughly a third, from 280 parts per million to 378 p.p.m. During the same period, concentrations of methane, an even more powerful (but more short-lived) greenhouse gas, have more than doubled, from .78 p.p.m. to 1.76 p.p.m. Scientists measure forcings in terms of watts per square meter, or w/m^2, by which they mean that a certain number of watts of energy have been added (or, in the case of a negative forcing, subtracted) for every single square meter of the earth's surface. The size of the greenhouse forcing is estimated, at this point, to be 2.5 w/m^2. A miniature Christmas light gives off about four tenths of a watt of energy, mostly in the form of heat, so that, in effect (as Sophie supposedly explained to Connor), we have covered the earth with tiny bulbs, six for every square meter. These bulbs are burning twenty-four hours a day, seven days a week, year in and year out.

If greenhouse gases were held constant at today's levels, it is estimated that it would take several decades for the full impact of the forcing that is already in place to be felt. This is because raising the earth's temperature involves not only warming the air and the surface of the land but also melting sea ice, liquefying glaciers, and, most significant, heating the oceans—all processes that require tremendous amounts of energy. (Imagine trying to thaw a gallon of ice cream or warm a pot of water using an Easy-Bake oven.) It could be argued that the delay that is built into the system is socially useful because it enables us—with the help of climate models—to prepare for what lies ahead, or that it is socially disastrous because it allows us to keep adding CO_2 to the atmosphere while fobbing the impacts off on our children and grandchildren. Either way, if current trends continue, which is to say, if steps are not taken to reduce emissions, carbon-dioxide levels will probably reach 500 parts per million—nearly double preindustrial levels—sometime around the middle of the century. By that point, of course, the forcing associated with greenhouse gases will also have increased, to four watts per square meter and possibly more. For comparison's sake, it is worth keeping in mind that the total forcing that ended the last ice age—a forcing that was eventually sufficient to melt mile-thick ice sheets and raise global sea levels by four hundred feet—is estimated to have been just six and a half watts per square meter.

There are two ways to operate a climate model. In the first, which is known as a transient run, greenhouse gases are slowly added to the simulated atmosphere—just as they would be to the real atmosphere—and the model forecasts what the effect of these additions will be at any given moment. In the second, greenhouse gases are added to the atmosphere all at once, and the model is run at these new levels until the climate has fully adjusted to the forcing by reaching a new equilibrium. Not surprisingly, this is known as an equilibrium run. For doubled CO_2, equilibrium runs of the GISS model predict that average global temperatures will rise by 4.9 degrees Fahrenheit. Only about a third of this increase is directly attributable to more greenhouse gases; the rest is a result of indirect effects, the most important among them being the so-called "water-vapor feedback." (Since warmer air holds more moisture, higher temperatures are expected to produce an atmosphere containing more water vapor, which is itself a greenhouse gas.) GISS's forecast is on the low end of the most recent projections; the Hadley Centre model, which is run by the British Met Office, predicts that for doubled CO_2 the eventual temperature rise will be 6.3 degrees Fahrenheit, while Japan's National Institute for Environmental Studies predicts 7.7 degrees.

In the context of ordinary life, a warming of 4.9, or even of 7.7, degrees may not seem like much to worry about; in the course of a normal summer's day, after all, air temperatures routinely rise by twenty degrees or more. Average global temperatures, however, have practically nothing to do with ordinary life. In the middle of the last glaciation, Manhattan, Boston, and Chicago were deep under ice, and sea levels were so low that Siberia and Alaska were connected by a land bridge nearly a thousand miles wide. At that point, average global temperatures were roughly ten degrees colder than they are today. Conversely, since our species evolved, average temperatures have never been much more than two or three degrees higher than they are right now.

This last point is one that climatologists find particularly significant. By studying Antarctic ice cores, researchers have been able to piece together a record both of the earth's temperature and of the composition of its atmosphere going back four full glacial cycles. (Temperature data can be extracted from the isotopic composition of the ice, and the makeup of the atmosphere can be reconstructed by analyzing tiny bubbles of trapped air.) What this record shows is that the planet is now nearly as warm as it has been at any point in the last four hundred and twenty thousand years. A possible consequence of even a four- or five-degree temperature rise—on the low end of projections for doubled CO_2—is that the world will enter a completely new climate regime, one with which modern humans have no prior experience. Meanwhile, at 378 p.p.m., CO_2 levels are significantly higher today than they have been at any other point in the Antarctic record. It is believed that the last time carbon-dioxide levels were in this range was three and a half million years ago, during what is known as the mid-Pliocene warm period, and they likely have not been much above it for tens of millions of years. A scientist with the National Oceanic and Atmospheric Administration (NOAA) put it to me—only half-jokingly—this way: "It's true that we've had higher CO_2 levels before. But, then, of course, we also had dinosaurs."

David Rind is a climate scientist who has worked at GISS since 1978. Rind acts as a trouble-shooter for the institute's model, scanning reams of numbers known as diagnostics, trying to catch problems, and he also works with GISS's Climate Impacts Group. (His office, like Hansen's, is filled with dusty piles of computer printouts.) Although higher temperatures are the most obvious and predictable result of increased CO_2, other, second-order consequences—rising sea levels, changes in vegetation, loss of snow cover—are likely to be just as significant. Rind's particular interest is how CO_2 levels will affect water supplies, because, as he put it to me, "you can't have a plastic version of water."

One afternoon, when I was talking to Rind in his office, he mentioned a visit that President Bush's science adviser, John Marburger, had paid to GISS a few years earlier. "He said, 'We're really interested in adaptation to climate change,' " Rind recalled. "Well, what does 'adaptation' mean?" He rummaged through one of his many file cabinets and finally pulled out a paper that he had published in the *Journal of Geophysical Research* entitled "Potential Evapo-transpiration and the Likelihood of Future Drought." In much the same way that wind velocity is measured using the Beaufort scale, water availability is measured using what's known as the Palmer Drought Severity Index. Different climate models offer very different predictions about future water availability; in the paper, Rind applied the criteria used in the Palmer index to GISS's model and also to a model operated by NOAA's Geophysical Fluid Dynamics Laboratory. He found that as carbon-dioxide levels rose, the world began to experience more and more serious water shortages, starting near the equator and then spreading toward the poles. When he applied the index to the GISS model for doubled CO_2, it showed most of the continental United States to be suffering under severe drought conditions. When he applied the index to the G.F.D.L. model, the results were even more dire. Rind created two maps to illustrate these findings. Yellow represented a 40 to 60 percent chance of summertime drought, ochre a

60 to 80 percent chance, and brown an 80 to 100 percent chance. In the first map, showing the GISS results, the Northeast was yellow, the Midwest was ochre, and the Rocky Mountain states and California were brown. In the second, showing the G.F.D.L. results, brown covered practically the entire country.

"I gave a talk based on these drought indices out in California to water-resource managers," Rind told me. "And they said, 'Well, if that happens, forget it.' There's just no way they could deal with that."

He went on, "Obviously, if you get drought indices like these, there's no adaptation that's possible. But let's say it's not that severe. What adaptation are we talking about? Adaptation in 2020? Adaptation in 2040? Adaptation in 2060? Because the way the models project this, as global warming gets going, once you've adapted to one decade you're going to have to change everything the next decade.

"We may say that we're more technologically able than earlier societies. But one thing about climate change is it's potentially geopolitically destabilizing. And we're not only more technologically able; we're more technologically able destructively as well. I think it's impossible to predict what will happen. I guess—though I won't be around to see it—I wouldn't be shocked to find out that by 2100 most things were destroyed." He paused. "That's sort of an extreme view."

ON THE OTHER SIDE OF THE HUDSON RIVER and slightly to the north of GISS, the Lamont-Doherty Earth Observatory occupies what was once a weekend estate in the town of Palisades, New York. The observatory is an outpost of Columbia University, and it houses, among its collections of natural artifacts, the world's largest assembly of ocean-sediment cores—more than thirteen thousand in all. The cores are kept in steel compartments that look like drawers from a filing cabinet, only longer and much skinnier. Some of the cores are chalky, some are clayey, and some are made up almost entirely of gravel. All can be coaxed to yield up—in one way or another—information about past climates.

Peter deMenocal is a paleoclimatologist who has worked at

Lamont-Doherty for fifteen years. He is an expert on ocean cores, and also on the climate of the Pliocene, which lasted from roughly five million to two million years ago. Around two and a half million years ago, the earth, which had been warm and relatively ice-free, started to cool down until it entered an era—the Pleistocene—of recurring glaciations. DeMenocal has argued that this transition was a key event in human evolution: right around the time that it occurred, at least two types of hominids—one of which would eventually give rise to us—branched off from a single ancestral line. Until quite recently, paleoclimatologists like deMenocal rarely bothered with anything much closer to the present day; the current interglacial—the Holocene—which began some ten thousand years ago, was believed to be, climatically speaking, too stable to warrant much study. In the mid-nineties, though, deMenocal, motivated by a growing concern over global warming—and a concomitant shift in government research funds—decided to look in detail at some Holocene cores. What he learned, as he put it to me when I visited him at Lamont-Doherty last fall, was "less boring than we had thought."

One way to extract climate data from ocean sediments is to examine the remains of what lived or, perhaps more pertinently, what died and was buried there. The oceans are rich with microscopic creatures known as foraminifera. There are about thirty planktonic species in all, and each thrives at a different temperature, so that by counting a species' prevalence in a given sample it is possible to estimate the ocean temperatures at the time the sediment was formed. When deMenocal used this technique to analyze cores that had been collected off the coast of Mauritania, he found that they contained evidence of recurring cool periods; every fifteen hundred years or so, water temperatures dropped for a few centuries before climbing back up again. (The most recent cool period corresponds to the Little Ice Age, which ended about a century and a half ago.) Also, perhaps even more significant, the cores showed profound changes in precipitation. Until about six thousand years ago, northern Africa was relatively wet—dotted with small lakes. Then it became dry, as it is today.

DeMenocal traced the shift to periodic variations in the earth's orbit, which, in a generic sense, are the same forces that trigger ice ages. But orbital changes occur gradually, over thousands of years, and northern Africa appears to have switched from wet to dry all of a sudden. Although no one knows exactly how this happened, it seems, like so many climate events, to have been a function of feedbacks—the less rain the continent got, the less vegetation there was to retain water, and so on until, finally, the system just flipped. The process provides yet more evidence of how a very small forcing sustained over time can produce dramatic results.

"We were kind of surprised by what we found," deMenocal told me about his work on the supposedly stable Holocene. "Actually, more than surprised. It was one of these things where, you know, in life you take certain things for granted, like your neighbor's not going to be an axe murderer. And then you discover your neighbor *is* an axe murderer."

NOT LONG AFTER DEMENOCAL BEGAN to think about the Holocene, a brief mention of his work on the climate of Africa appeared in a book produced by *National Geographic*. On the facing page, there was a piece on Harvey Weiss and his work at Tell Leilan. DeMenocal vividly remembers his reaction. "I thought, Holy cow, that's just amazing!" he told me. "It was one of these cases where I lost sleep that night, I just thought it was such a cool idea."

DeMenocal also recalls his subsequent dismay when he went to learn more. "It struck me that they were calling on this climate-change argument, and I wondered how come I didn't know about it," he said. He looked at the *Science* paper in which Weiss had originally laid out his theory. "First of all, I scanned the list of authors and there was no paleoclimatologist on there," deMenocal said. "So then I started reading through the paper and there basically was no paleoclimatology in it." (The main piece of evidence Weiss adduced for a drought was that Tell Leilan had filled with dust.) The more deMenocal thought about it, the more unconvincing he found the data, on the one hand, and the more compelling he found the underlying idea, on the other. "I just couldn't leave it alone," he told me. In the summer of 1995, he went with Weiss to Syria to visit Tell Leilan. Subsequently, he decided to do his own study to prove—or disprove—Weiss's theory.

Instead of looking in, or even near, the ruined city, deMenocal focused on the Gulf of Oman, nearly a thousand miles downwind. Dust from the Mesopotamian floodplains, just north of Tell Leilan, contains heavy concentrations of the mineral dolomite, and since arid soil produces more wind-borne dust, deMenocal figured that if there had been a drought of any magnitude it would show up in gulf sediments. "In a wet period, you'd be getting none or very, very low amounts of dolomite, and during a dry period you'd be getting a lot," he explained. He and a graduate student named Heidi Cullen developed a highly sensitive test to detect dolomite, and then Cullen assayed, centimeter by centimeter, a sediment core that had been extracted near where the Gulf of Oman meets the Arabian Sea.

"She started going up through the core," deMenocal told me. "It was like nothing, nothing, nothing, nothing, nothing. Then one day, I think it was a Friday afternoon, she goes, 'Oh my God.' It was really classic." DeMenocal had thought that the dolomite level, if it was elevated at all, would be modestly higher; instead, it went up by 400 percent. Still, he wasn't satisfied. He decided to have the core re-analyzed using a different marker: the ratio of strontium 86 and strontium 87 isotopes. The same spike showed up. When deMenocal had the core carbon-dated, it turned out that the spike lined up exactly with the period of Tell Leilan's abandonment.

Tell Leilan was never an easy place to live. Much like, say, western Kansas today, the Khabur plains received enough annual rainfall—about seventeen inches—to support cereal crops but not enough to grow much else. "Year-to-year variations were a real threat, and so they obviously needed to have grain storage and to have ways to buffer themselves," deMenocal observed. "One generation would tell the next, 'Look, there are these things that happen that you've got to be prepared for.' And they were good at that. They could manage that. They were there for hundreds of years."

He went on, "The thing they couldn't prepare for was the same thing that we won't prepare for, because in their case they didn't know about it and because in our case the political system can't listen to it. And that is that the climate system has much greater things in store for us than we think."

Shortly before Christmas, Harvey Weiss gave a lunchtime lecture at Yale's Institute for Biospheric Studies. The title was "What Happened in the Holocene," which, as Weiss explained, was an allusion to a famous archaeology text by V. Gordon Childe, entitled "What Happened in History." The talk brought together archaeological and paleoclimatic records from the Near East over the last ten thousand years.

Weiss, who is sixty years old, has thinning gray hair, wire-rimmed glasses, and an excitable manner. He had prepared for the audience—mostly Yale professors and graduate students—a handout with a time line of Mesopotamian history. Key cultural events appeared in black ink, key climatological ones in red. The two alternated in a rhythmic cycle of disaster and innovation. Around 6200 BC, a severe global cold snap—red ink—produced aridity in the Near East. (The cause of the cold snap is believed to have been a catastrophic flood that emptied an enormous glacial lake—called Lake Agassiz—into the North Atlantic.) Right around the same time—black ink—farming villages in northern Mesopotamia were abandoned, while in central and southern Mesopotamia the art of irrigation was invented. Three thousand years later, there was another cold snap, after which settlements in northern Mesopotamia once again were deserted. The most recent red event, in 2200 BC, was followed by the dissolution of the Old Kingdom in Egypt, the abandonment of villages in ancient Palestine, and the fall of Akkad. Toward the end of his talk, Weiss, using a PowerPoint program, displayed some photographs from the excavation of Tell

Leilan. One showed the wall of a building—probably intended for administrative offices—that had been under construction when the rain stopped. The wall was made from blocks of basalt topped by rows of mud bricks. The bricks gave out abruptly, as if construction had ceased from one day to the next.

The monochromatic sort of history that most of us grew up with did not allow for events like the drought that destroyed Tell Leilan. Civilizations fell, we were taught, because of wars or barbarian invasions or political unrest. (Another famous text by Childe bears the exemplary title "Man Makes Himself.") Adding red to the time line points up the deep contingency of the whole enterprise. Civilization goes back, at the most, ten thousand years, even though, evolutionarily speaking, modern man has been around for at least ten times that long. The climate of the Holocene was not boring, but at least it was dull enough to allow people to sit still. It is only after the immense climatic shifts of the glacial epoch had run their course that writing and agriculture finally emerged.

Nowhere else does the archaeological record go back so far or in such detail as in the Near East. But similar red-and-black chronologies can now be drawn up for many other parts of the world: the Indus Valley, where, some four thousand years ago, the Harappan civilization suffered a decline after a change in monsoon patterns; the Andes, where, fourteen hundred years ago, the Moche abandoned their cities in a period of diminished rainfall; and even the United States, where the arrival of the English colonists on Roanoke Island, in 1587, coincided with a severe regional drought. (By the time English ships returned to resupply the colonists, three years later, no one was left.) At the height of the Mayan civilization, population density was five hundred per square mile, higher than it is in most parts of the United States today. Two hundred years later, much of the territory occupied by the Mayans had been completely depopulated. You can argue that man through culture creates stability, or you can argue, just as plausibly, that stability is for culture an essential precondition.

After the lecture, I walked with Weiss back to his office, which is near the center of the Yale campus, in the Hall of Graduate Studies. This past year, Weiss decided to suspend excavation at Tell Leilan. The site lies only fifty miles from the Iraqi border, and, owing to the uncertainties of the war, it seemed like the wrong sort of place to bring graduate students. When I visited, Weiss had just returned from a trip to Damascus, where he had gone to pay the guards who watch over the site when he isn't there. While he was away from his office, its contents had been piled up in a corner by repairmen who had come to fix some pipes. Weiss considered the piles disconsolately, then unlocked a door at the back of the room.

The door led to a second room, much larger than the first. It was set up like a library, except that instead of books the shelves were stacked with hundreds of cardboard boxes. Each box contained fragments of broken pottery from Tell Leilan. Some were painted, others were incised with intricate designs, and still others were barely distinguishable from pebbles. Every fragment had been inscribed with a number, indicating its provenance.

I asked what he thought life in Tell Leilan had been like. Weiss told me that that was a "corny question," so I asked him about the city's abandonment. "Nothing allows you to go beyond the third or fourth year of a drought, and by the fifth or sixth year you're probably gone," he observed. "You've given up hope for the rain, which is exactly what they wrote in 'The Curse of Akkad.' " I asked to see something that might have been used in Tell Leilan's last days. Swearing softly, Weiss searched through the rows until he finally found one particular box. It held several potsherds that appeared to have come from identical bowls. They were made from a greenish-colored clay, had been thrown on a wheel, and had no decoration. Intact, the bowls had held about a liter, and Weiss explained that they had been used to mete out rations—probably wheat or barley—to the workers of Tell Leilan. He passed me one of the fragments. I held it in my hand for a moment and tried to imagine the last Akkadian who had touched it. Then I passed it back.

DENNIS OVERBYE

Remembrance of Things Future: The Mystery of Time

FROM THE *NEW YORK TIMES*

In May of 2005, a graduate student at MIT organized the first-ever convention for time travelers. While the odds were slight that any time traveler might actually appear, the disturbing reality is that the laws of physics do not prevent time travel. Dennis Overbye explores just what physicists make of this peculiarity.

There was a conference for time travelers at MIT earlier this spring.

I'm still hoping to attend, and although the odds are slim, they are apparently not zero despite the efforts and hopes of deterministically minded physicists who would like to eliminate the possibility of your creating a paradox by going back in time and killing your grandfather.

"No law of physics that we know of prohibits time travel," said Dr. J. Richard Gott, a Princeton astrophysicist.

Dr. Gott, author of the 2001 book *Time Travel in Einstein's Universe: The Physical Possibilities of Travel Through Time*, is one of a small breed of physicists who spend part of their time (and their research grants) thinking about wormholes in space, warp drives, and other cosmic constructions that "absurdly advanced civilizations" might use to travel through time.

It's not that physicists expect to be able to go back and attend Woodstock, drop by the Bern patent office to take Einstein to lunch, see the dinosaurs; or investigate John F. Kennedy's assassination.

In fact, they're pretty sure those are absurd dreams and are all bemused by the fact that they can't say why. They hope such extreme theorizing could reveal new features, gaps or perhaps paradoxes or contradictions in the foundations of Physics As We Know It and point the way to new ideas.

"Traversable wormholes are primarily useful as a 'gedanken experiment' to explore the limitations of general relativity," said Dr. Francisco Lobo of the University of Lisbon.

If general relativity, Einstein's theory of gravity and space-time, allows for the ability to go back in time and kill your grandfather, asks Dr. David Z. Albert, a physicist and philosopher at Columbia University, "how can it be a logically consistent theory?"

In his recent book *The Universe in a Nutshell,* Dr. Stephen W. Hawking wrote, "Even if it turns out that time travel is impossible, it is important that we understand why it is impossible."

When it comes to the nature of time, physicists are pretty much at as much of a loss as the rest of us who seem hopelessly swept along in its current. The mystery of time is connected with some of the thorniest questions in physics, as well as in philosophy, like why we remember the past but not the future, how causality works, why you can't stir cream out of your coffee or put perfume back in a bottle.

But some theorists think that has to change.

Just as Einstein needed to come up with a new concept of time in order to invent relativity one hundred years ago this year, so physicists say that a new insight into time—or beyond it—may be required to crack profound problems like how the universe began, what happens at the center of a black hole, or how to marry relativity and quantum theory into a unified theory of nature.

Space and time, some quantum gravity theorists say, are most likely a sort of illusion—or less sensationally, an "approximation"—doomed to be replaced by some more fundamental idea. If only they could think of what that idea is.

"By convention there is space, by convention time," Dr. David J. Gross, director of the Kavli Institute for Theoretical Physics and a winner of last year's Nobel Prize, said recently, paraphrasing the Greek philosopher Democritus, "in reality there is . . . ?" his voice trailing off.

The issues raised by time travel are connected to these questions, Dr. Lawrence Krauss, a physicist at Case Western Reserve University in Cleveland and author of the book *The Physics of Star Trek,* said. "The minute you have time travel you have paradoxes," Dr. Krauss said, explaining that if you can go backward in time, you confront fundamental issues like cause and effect or the meaning of your own identity if there can be two of you at once. A refined theory of time would have to explain "how a sensible world could result from something so nonsensical."

"That's why time travel is philosophically important and has captivated the public, who care about these paradoxes," he said.

At stake, said Dr. Albert, the philosopher and author of his own time book, *Time and Chance,* is "what kind of view science presents us of the world.

"Physics gets time wrong, and time is the most familiar thing there is," Dr. Albert said.

We all feel time passing in our bones, but ever since Galileo and Newton in the seventeenth century began using time as a coordinate to help chart the motion of cannonballs, time—for physicists—has simply been an "addendum in the address of an event," Dr. Albert said.

"There is a feeling in philosophy," he said, "that this picture leaves no room for locutions about flow and the passage of time we experience."

Then there is what physicists call "the arrow of time" problem. The fundamental laws of physics don't care what direction time goes, he pointed out. Run a movie of billiard balls colliding or planets swirling around in their orbits in reverse and nothing will look weird, but if you run a movie of a baseball game in reverse people will laugh.

Einstein once termed the distinction between past, present, and future "a stubborn illusion," but as Dr. Albert said, "It's hard to imagine something more basic than the distinction between the future and the past."

SPACE AND TIME, the philosopher Augustine famously argued seventeen hundred years ago, are creatures of existence and the universe, born with it, not separately standing features of eternity. That is the same answer that Einstein came up with in 1915 when he finished his general theory of relativity.

That theory explains how matter and energy warp the geometry of space and time to produce the effect we call gravity. It also predicted, somewhat to Einstein's dismay, the expansion of the universe, which forms the basis of modern cosmology.

But Einstein's theory is incompatible, mathematically and philosophically, with the quirky rules known as quantum mechanics that describe the microscopic randomness that fills this elegantly curved expanding space-time. According to relativity, nature is continuous, smooth and orderly; in quantum theory the world is jumpy and discontinuous. The sacred laws of physics are correct only on average.

Until the pair are married in a theory of so-called quantum gravity, physics has no way to investigate what happens in the Big Bang, when the entire universe is so small that quantum rules apply.

Looked at closely enough, with an imaginary microscope that could see lengths down to 10^{-33} centimeters, quantum gravity theorists say, even ordinary space and time dissolve into a boiling mess that Dr. John Wheeler, the Princeton physicist and phrasemaker, called "space-time foam." At that level of reality, which exists underneath all our fingernails, clocks and rulers as we know them cease to exist.

"Everything we know about stops at the Big Bang, the Big Crunch," said Dr. Raphael Bousso, a physicist at the University of California, Berkeley.

What happens to time at this level of reality is anybody's guess. Dr. Lee Smolin, of the Perimeter Institute for Theoretical Physics in Waterloo, Ontario, said, "There are several different, very different, ideas about time in quantum gravity."

One view, he explained, is that space and time "emerge" from this foamy substrate when it is viewed at larger scales. Another is that space emerges but that time or some deeper relations of cause and effect are fundamental.

Dr. Fotini Markopoulou Kalamara of the Perimeter Institute described time as, if not an illusion, an approximation, "a bit like the way you can see the river flow in a smooth way even though the individual water molecules follow much more complicated patterns."

She added in an e-mail message: "I have always thought that there has to be some basic fundamental notion of causality, even if it doesn't look at all like the one of the space-time we live in. I can't see how to get causality from something that has none; neither have I ever seen anyone succeed in doing so."

Physicists say they have a sense of how space can emerge, because of recent advances in string theory, the putative theory of everything, which posits that nature is composed of wriggling little strings.

Calculations by Dr. Juan Maldacena of the Institute for Advanced Study in Princeton and by others have shown how an extra dimension of space can pop mathematically into being almost like magic, the way the illusion of three dimensions can appear in the holograms on bank cards. But string theorists admit they don't know how to do the same thing for time yet.

"Time is really difficult," said Dr. Cumrun Vafa, a Harvard string theorist. "We have not made much progress on the emergence of time. Once we make progress we will make progress on the early universe, on high energy physics and black holes.

"We are out on a limb trying to understand what's going on here."

Dr. Bousso, an expert on holographic theories of space-time, said that in general relativity time gets no special treatment.

He said he expected both time and space to break down, adding, "We really just don't know what's going to go.

"There is a lot of mysticism about time," Dr. Bousso said. "Time is what a clock measures. What a clock measures is more interesting than you thought."

"IF WE COULD GO FASTER THAN LIGHT, we could telegraph into the past," Einstein once said. According to the theory of special relativity—which he proposed in 1905 and which ushered $E=mc^2$ into the world and set the speed of light as the cosmic speed limit—such telegraphy is not possible, and there is no way of getting back to the past.

But, somewhat to Einstein's surprise, in general relativity it is possible to beat a light beam across space. That theory, which Einstein finished in 1916, said that gravity resulted from the warping of space-time geometry by matter and energy, the way a bowling ball sags a trampoline. And all this warping and sagging can create shortcuts through space-time.

In 1949, Kurt Gödel, the Austrian logician and mathematician then at the Institute for Advanced Study, showed that in a rotating universe, according to general relativity, there were paths, technically called "closed timelike curves," you could follow to get back to the past. But it has turned out that the universe does not rotate very much, if at all.

Most scientists, including Einstein, resisted the idea of time travel until 1988 when Dr. Kip Thorne, a gravitational theorist at the California Institute of Technology, and two of his graduate students, Dr. Mike Morris and Dr. Ulvi Yurtsever, published a pair of papers concluding that the laws of physics may allow you to use wormholes, which are like tunnels through space connecting distant points, to travel in time.

These holes, technically called Einstein-Rosen bridges, have long been predicted as a solution of Einstein's equations. But physicists dismissed them because calculations predicted that gravity would slam them shut.

Dr. Thorne was inspired by his friend, the late Cornell scientist and author Carl Sagan, who was writing the science fiction novel *Contact*, later made into a Jodie Foster movie, and was looking for a way to send his heroine, Eleanor Arroway, across the galaxy. Dr. Thorne and his colleagues imagined that such holes could be kept from collapsing and thus maintained to be used as a galactic subway, at least in principle, by threading them with something called Casimir energy (after the Dutch physicist Hendrik Casimir), which is a sort of quantum suction produced when two parallel metal plates are placed very close together. According to Einstein's equations, this suction, or negative pressure, would have an antigravitational effect, keeping the walls of the wormhole apart.

If one mouth of a wormhole was then grabbed by a spaceship and taken on a high-speed trip, according to relativity, its clock would run slow compared with the other end of the wormhole. So the wormhole would become a portal between two different times as well as places.

Dr. Thorne later said he had been afraid that the words "time travel" in the second paper's title would create a sensation and tarnish his students' careers, and he had forbidden Caltech to publicize it.

In fact, their paper made time travel safe for serious scientists, and other theorists, including Dr. Frank Tipler of Tulane University and Dr. Hawking, jumped in. In 1991, for example, Dr. Gott of Princeton showed how another shortcut through space-time could be manufactured using pairs of cosmic strings—dense tubes of primordial energy not to be confused with the strings of string theory, left over by the Big Bang in some theories of cosmic evolution—rushing past each other and warping space around them.

THESE SPECULATIONS HAVE BEEN BOLSTERED (not that time machine architects lack imagination) by the unsettling discovery that the universe may be full of exactly the kind of antigravity stuff needed to grow and prop open a wormhole. Some mysterious "dark energy," astronomers say, is pushing space apart and accelerating the expansion of the universe. The race is on to measure this energy precisely and find out what it is.

Among the weirder and more disturbing explanations for this cosmic riddle is something called phantom energy, which is so virulently antigravitational that it would eventually rip planets, people, and even atoms apart, ending everything. As it happens, this bizarre stuff would also be perfect for propping open a wormhole, Dr. Lobo of Lisbon recently pointed out. "This certainly is an interesting prospect for an absurdly advanced civilization, as phantom energy probably comprises seventy percent of the universe," Dr. Lobo wrote in an e-mail message. Dr. Sergey Sushkov of Kazan State Pedagogical University in Russia has made the same suggestion.

In a paper posted on the physics Web site arxiv.org/abs/gr-qc/0502099, Dr. Lobo suggested that as the universe was stretched and stretched under phantom energy, microscopic holes in the quantum "space-time foam" might grow to macroscopic usable size. "One could also imagine an advanced civilization mining the cosmic fluid for phantom energy necessary to construct and sustain a traversable wormhole," he wrote.

Such a wormhole, he even speculated, could be used to escape the "big rip" in which a phantom energy universe will eventually end.

But nobody knows if phantom, or exotic, energy is really allowed in nature and most physicists would be happy if it is not. Its existence would lead to paradoxes, like negative kinetic energy, where something could lose energy by speeding up, violating what is left of common sense in modern physics.

Dr. Krauss said, "From the point of view of realistic theories, phantom energy just doesn't exist."

But such exotic stuff is not required for all time machines, Dr. Gott's cosmic strings for example. In another recent paper, Dr. Amos Ori of the Technion-Israel Institute of Technology in Haifa describes a time machine that he claims can be built by moving around colossal masses to warp the space inside a doughnut of regular empty space into a particular configuration, something an advanced civilization may be able to do in one hundred or two hundred years.

The space inside the doughnut, he said, will then naturally evolve according to Einstein's laws into a time machine.

Dr. Ori admits that he doesn't know if his machine would be stable. Time machines could blow up as soon as you turned them on, say some physicists, including Dr. Hawking, who has proposed what he calls the "chronology protection" conjecture to keep the past safe for historians. Random microscopic fluctuations in matter and energy and space itself, they argue, would be amplified by going around and around boundaries of the machine or the wormhole, and finally blow it up.

Dr. Gott and his colleague Dr. Li-Xin Li have shown that there are at least some cases where the time machine does not blow up. But until gravity marries quantum theory, they admit, nobody knows how to predict exactly what the fluctuations would be.

"That's why we really need to know about quantum gravity," Dr. Gott said. "That's one reason people are interested in time travel."

BUT WHAT ABOUT KILLING YOUR GRANDFATHER? In a well-ordered universe, that would be a paradox and shouldn't be able to happen, everybody agrees.

That was the challenge that Dr. Joe Polchinski, now at the Kavli Institute for Theoretical Physics in Santa Barbara, California, issued to Dr. Thorne and his colleagues after their paper was published.

Being a good physicist, Dr. Polchinski phrased the problem in terms of billiard balls. A billiard ball, he suggested, could roll into one end of a time machine, come back out the other end a little earlier and collide with its earlier self, thereby preventing itself from entering the time machine to begin with.

Dr. Thorne and two students, Fernando Echeverria and Gunnar Klinkhammer, concluded after months of mathematical struggle that there was a logically consistent solution to the billiard matricide that Dr. Polchinski had set up. The ball would come back out of the time machine and deliver only a glancing blow to itself, altering its path just enough so that it would still hit the time machine. When it came back out, it would be aimed just so as to deflect itself rather than hitting full on. And so it would go like a movie with a circular plot.

In other words, it's not a paradox if you go back in time and save your grandfather. And, added Dr. Polchinski, "It's not a paradox if you try to shoot your grandfather and miss."

"The conclusion is somewhat satisfying," Dr. Thorne wrote in his book *Black Holes and Time Warps: Einstein's Outrageous Legacy.* "It suggests that the laws of physics might accommodate themselves to time machines fairly nicely."

Dr. Polchinski agreed. "I was making the point that the grandfather paradox had nothing to do with free will, and they found a nifty resolution," he said in an e-mail message, adding, nevertheless, that his intuition still tells him time machines would lead to paradoxes.

Dr. Bousso said, "Most of us would consider it quite satisfactory if the laws of quantum gravity forbid time travel."

W. Wayt Gibbs

Obesity: An Overblown Epidemic?

FROM *SCIENTIFIC AMERICAN*

In the past few years, scientists have made headlines by announcing that too many Americans are perilously overweight, making them vulnerable to disease and shortening their lives. But a growing number of researchers are questioning these findings. As W. Wayt Gibbs shows, the evidence for an "obesity epidemic" is surprisingly thin.

Could it be that excess fat is not, by itself, a serious health risk for the vast majority of people who are overweight or obese—categories that in the United States include about six of every ten adults? Is it possible that urging the overweight or mildly obese to cut calories and lose weight may actually do more harm than good?

Such notions defy conventional wisdom that excess adiposity kills more than three hundred thousand Americans a year and that the gradual fattening of nations since the 1980s presages coming epidemics of diabetes, cardiovascular disease, cancer, and a host of other medical consequences. Indeed, just this past March the *New England Journal of Medicine* presented a "Special Report," by S. Jay Olshansky, David B. Allison, and others that seemed to confirm such fears. The authors asserted that because of the obesity epidemic, "the steady rise in life expectancy during the past two centuries may soon come to an end." Articles about the special report by the *New York Times*, the *Washington Post*, and many other news outlets emphasized its forecast that obesity may shave up to five years off average life spans in coming decades.

And yet an increasing number of scholars have begun accusing obesity experts, public health officials, and the media of exaggerating the health effects of the epidemic of overweight and obesity. The charges appear in a recent flurry of scholarly books, including *The Obesity Myth* by Paul F. Campos (Gotham Books, 2004); *The Obesity Epidemic: Science, Morality and Ideology* by Michael Gard and Jan Wright (Routledge, 2005); *Obesity: The Making of an American Epidemic* by J. Eric Oliver (Oxford University Press, August 2005); and a book on popular misconceptions about diet and weight gain by Barry Glassner (to be published in 2007 by HarperCollins).

These critics, all academic researchers outside the medical community, do not dispute surveys that find the obese fraction of the population to have roughly doubled in the United States and many parts of Europe since 1980. And they acknowledge that obesity, especially in its extreme forms, does seem to be a factor in some illnesses and premature deaths.

They allege, however, that experts are blowing hot air when they warn that overweight and obesity are causing a massive, and worsening, health crisis. They scoff, for example, at the 2003 assertion by Julie L. Gerberding, director of the Centers for Disease Control and Prevention, that "if you looked at any epidemic—whether it's influenza or plague from the Middle Ages—they are not as serious as the epidemic of obesity in terms of the health impact on our country and our society." (An epidemic of influenza killed forty million people world-wide between 1918 and 1919, including 675,000 in the United States.)

What is really going on, asserts Oliver, a political scientist at the University of Chicago, is that "a relatively small group of scientists and doctors, many directly funded by the weight-loss industry, have created an arbitrary and unscientific definition of overweight and obesity. They have inflated claims and distorted statistics on the consequences of our growing weights, and they have largely ignored the complicated health realities associated with being fat."

One of those complicated realities, concurs Campos, a professor of law at the University of Colorado at Boulder, is the widely accepted evidence that genetic differences account for 50 to 80 percent of the variation in fatness within a population. Because no safe and widely practical methods have been shown to induce long-term loss of more than about 5 percent of body weight, Campos says, "health authorities are giving people advice—maintain a body mass index in the 'healthy weight' range—that is literally impossible for many of them to follow." Body mass index, or BMI, is a weight-to-height ratio.

By exaggerating the risks of fat and the feasibility of weight loss, Campos and Oliver claim, the CDC, the U.S. Department of Health and Human Services and the World Health Organization inadvertently perpetuate stigma, encourage unbalanced diets, and, perhaps, even exacerbate weight gain. "The most perverse irony is that we may be creating a disease simply by labeling it as such," Campos states.

On first hearing, these dissenting arguments may sound like nonsense. "If you really look at the medical literature and think obesity isn't bad, I don't know what planet you are on," says James O. Hill, an obesity researcher at the University of Colorado Health Sciences Center. New dietary guidelines issued by the DHHS and the U.S. Department of Agriculture in January state confidently that "a high prevalence of overweight and obesity is of great public health concern because excess body fat leads to a higher risk for premature death, type 2 diabetes, hypertension, dyslipidemia [high cholesterol], cardiovascular disease, stroke, gall bladder disease, respiratory dysfunction, gout, osteoarthritis, and certain kinds of cancers." The clear implication is that any degree of overweight is dangerous and that a high BMI is not merely a marker of high risk but a cause.

"These supposed adverse health consequences of being 'overweight' are not only exaggerated but for the most part are simply fabricated," Campos alleges. Surprisingly, a careful look at recent epidemiological studies and clinical trials suggests that the critics, though perhaps overstating some of their accusations, may be onto something.

Oliver points to a new and unusually thorough analysis of three large, nationally representative surveys, for example, that found only a very slight—and statistically insignificant—increase in mortality among mildly obese people, as compared with those in the "healthy weight" category, after subtracting the effects of age, race, sex, smoking and alcohol consumption. The three surveys—medical measurements collected in the early 1970s, late 1970s, and early 1990s, with subjects matched against death registries nine to nineteen years later—indicate that it is much more likely that U.S. adults who fall in the overweight category have a *lower* risk of premature death than do those of so-called healthy weight. The overweight segment of the "epidemic of overweight and obesity" is more likely reducing death rates than boosting them. "The majority of Americans who weigh too much are in this category," Campos notes.

Counterintuitively, "underweight, even though it occurs in only a tiny fraction of the population, is actually associated with more excess deaths than class I obesity," says Katherine M. Flegal, a senior research scientist at the CDC. Flegal led the study, which appeared in the *Journal of the Amercian Medical Association* on April 20 after undergoing four months of scrutiny by internal reviewers at the CDC and the National Cancer Institute and additional peer review by the journal.

These new results contradict two previous estimates that were the basis of the oft-repeated claim that obesity cuts short three hundred thousand or more lives a year in the United States. There are good reasons to suspect, however, that both these earlier estimates were compromised by dubious assumptions, statistical errors, and outdated measurements.

When Flegal and her coworkers analyzed just the most recent survey, which measured heights and weights from 1988 to 1994 and deaths up to 2000, even severe obesity failed to show up as a statistically significant mortality risk. It seems probable, Flegal speculates, that in recent decades improvements in medical care have reduced the mortality level associated with obesity. That would square, she observes, with both the unbroken rise in life expectancies and the uninterrupted fall in death rates attributed to heart disease and stroke throughout the entire twenty-five-year spike in obesity in the United States.

But what about the warning by Olshansky and Allison that the toll from obesity is yet to be paid, in the form of two to five years of life lost? "These are just back-of-the-envelope, plausible scenarios," Allison hedges, when pressed. "We never meant for them to be portrayed as precise." Although most media reports jumped on the "two to five years" quote, very few mentioned that the paper offered no statistical analysis to back it up.

The life expectancy costs of obesity that Olshansky and his colleagues actually calculated were based on a handful of convenient, but false, presuppositions. First, they assumed that every obese American adult currently has a BMI of 30, or alternatively of 35—the upper and lower limits of the "mild obesity" range. They then compared that simplified picture of the United States with an imagined nation in which no adult has a BMI of more than 24—the upper limit of "healthy weight"—and in which underweight causes zero excess deaths.

To project death rates resulting from obesity, the study used risk data that are more than a decade old rather than the newer ratios Flegal included, which better reflect dramatically improved treatments for cardiovascular disease and diabetes. The authors further assumed not only that the old mortality risks have remained constant but also that future advances in medicine will have no effect whatsoever on the health risks of obesity.

If all these simplifications are reasonable, the March paper concluded, then the estimated hit to the average life expectancy of the U.S. population from its world-leading levels of obesity is four to nine months. ("Two to five years" was simply a gloomy guess of what could happen in "coming decades" if an increase in overweight children was to fuel additional spikes in adult obesity.) The study did not attempt to determine whether, given its many uncertainties, the number of months lost was reliably different from zero. Yet in multiple television and newspaper interviews about the study, co-author David S. Ludwig evinced full confidence as he compared the effect of rising obesity rates to "a massive tsunami headed toward the United States."

Critics decry episodes such as this one as egregious examples of a general bias in the obesity research community. Medical researchers tend to cast the expansion of waistlines as an impending disaster "because it inflates their stature and allows them to get more research grants. Government health agencies wield it as a rationale for their budget allocations," Oliver writes. (The National Institutes of Health increased its funding for obesity research by 10 percent in 2005, to $440 million.) "Weight-loss companies and surgeons employ it to get their services covered by insurance," he continues. "And the pharmaceutical industry uses it to justify new drugs."

"The war on fat," Campos concurs, "is really about making some of us rich." He points to the financial support that many influential obesity researchers receive from the drug and diet industries. Allison, a professor at the University of Alabama at Birmingham, discloses payments from 148 such companies, and Hill says he has consulted with some of them as well. (Federal policies prohibit Flegal and other CDC scientists from accepting non-governmental wages.) None of the dissenting authors cites evidence of anything more than a potential conflict of interest, however.

EVEN THE BEST MORTALITY STUDIES provide only a flawed and incomplete picture of the health consequences of the obesity epidemic, for three reasons. First, by counting all lives lost to obesity, the studies so far have ignored the fact that some diversity in human body size is normal and that every well-nourished population thus contains some obese people. The "epidemic" refers to a sudden increase in obesity, not its mere existence. A proper accounting of the epidemic's mortal cost would estimate only the number of lives cut short by whatever amount of obesity exceeds the norm.

Second, the analyses use body mass index as a convenient proxy for body fat. But BMI is not an especially reliable stand-in. And third, although everyone cares about mortality, it is not the only thing that we care about. Illness and quality of life matter a great deal, too.

All can agree that severe obesity greatly increases the risk of numerous diseases, but that form of obesity, in which BMI exceeds 40, affects only about one in twelve of the roughly 130 million American adults who set scales spinning above the "healthy" range. At issue is whether rising levels of overweight, or of mild to moderate obesity, are pulling up the national burden of heart disease, cancer, and diabetes.

In the case of heart disease, the answer appears to be no—or at least not yet. U.S. health agencies do not collect annual figures on the incidence of cardiovascular disease, so researchers look instead for trends in mortality and risk factors, as measured in periodic surveys. Both have been falling.

Alongside Flegal's April paper in *JAMA* was another by Edward W. Gregg and his colleagues from the CDC that found that in the United States the prevalence of high blood pressure dropped by half between 1960 and 2000. High cholesterol followed the same trend—and both declined more steeply among the overweight and obese than among those of healthy weight. So although high blood pressure is still twice as common among the obese as it is among the lean, the paper notes that "obese persons now have better [cardiovascular disease] risk profiles than their leaner counterparts did twenty to thirty years ago."

The new findings reinforce those published in 2001 by a ten-year WHO study that examined 140,000 people in thirty-eight cities on four continents. The investigators, led by Alun Evans of the Queen's University of Belfast, saw broad increases in BMI and equally broad declines in high blood pressure and high cholesterol. "These facts are hard to reconcile," they wrote.

It may be, Gregg suggests, that better diagnosis and treatment of high cholesterol and blood pressure have more than compensated for any increases from rising obesity. It could also be, he adds, that obese people are getting more exercise than they used to; regular physical activity is thought to be a powerful preventative against heart disease.

Oliver and Campos explore another possibility: that fatness is partially—or even merely—a visible marker of other factors that are more important but harder to perceive. Diet composition, physical fitness, stress levels, income, family history and the location of fat within the body are just a few of one-hundred-odd "independent" risk factors for cardiovascular disease identified in the medical literature. The observational studies that link obesity to heart disease ignore nearly all of them and in doing so effectively assign their causal roles to obesity. "By the same criteria we are blaming obesity for heart disease," Oliver writes, "we could accuse smelly clothes, yellow teeth, or bad breath for lung cancer instead of cigarettes."

As for cancer, a 2003 report on a sixteen-year study of 900,000 American adults found significantly increased death rates for several kinds of tumors among overweight or mildly obese people. Most of these apparently obesity-related cancers are very rare, however, killing at most a few dozen people a year for every 100,000 study participants. Among women with a high BMI, both colon cancer and post-menopausal breast cancer risks were slightly elevated; for overweight and obese men, colon and prostate cancer presented the most common increased risks. For both women and men, though, being overweight or obese seemed to confer significant *protection* against lung cancer, which is by far the most commonly lethal malignancy. That relation held even after the effects of smoking were subtracted.

Obesity's Catch-22

It is through type 2 diabetes that obesity seems to pose the biggest threat to public health. Doctors have found biological connections between fat, insulin, and the high blood sugar levels that define the disease. The CDC estimates that 55 percent of adult diabetics are obese, significantly more than the 31 percent prevalence of obesity in the general population. And as obesity has become more common, so, too, has diabetes, suggesting that one may cause the other.

Yet the critics dispute claims that diabetes is soaring (even among children), that obesity is the cause, and that weight loss is the solution. A 2003 analysis by the CDC found that "the prevalence of diabetes, either diagnosed or undiagnosed, and of impaired fasting glucose did not appear to increase substantially during the 1990s," despite the sharp rise in obesity.

"Undiagnosed diabetes" refers to people who have a single positive test for high blood sugar in the CDC surveys. (Two or more positive results are required for a diagnosis of diabetes.) Gregg's paper in April reiterates the oft-repeated "fact" that for every five adults diagnosed with diabetes, there are three more diabetics who are undiagnosed. "Suspected diabetes" would be a better term, however, because the single test used by the CDC may be wildly unreliable.

In 2001 a French study of 5,400 men reported that 42 percent of the men who tested positive for diabetes using the CDC method turned out to be nondiabetic when checked by a "gold standard" test thirty months later. The false negative rate—true diabetics missed by the single blood test—was just 2 percent.

But consider the growing weights of children, Hill urges. "You're getting kids at ten to twelve years of age developing type 2 diabetes. Two generations ago you never saw a kid with it."

Anecdotal evidence often misleads, Campos responds. He notes that when CDC researchers examined 2,867 adolescents in the NHANES survey of 1988 to 1994, they identified just four that had type 2 diabetes. A more focused study in 2003 looked at 710 "grossly obese" boys and girls ages six to eighteen in Italy. These kids were the heaviest of the heavy, and more than half had a family history (and thus an inherited risk) of diabetes. Yet only one of the 710 had type 2 diabetes.

Nevertheless, as many as 4 percent of U.S. adults might have diabetes because of their obesity—if fat is in fact the most important cause of the disease. "But it may be that type 2 diabetes causes fatness," Campos argues. (Weight gain is a common side effect of many diabetes drugs.) "A third factor could cause both type 2 diabetes and fatness." Or it could be some complex combination of all these, he speculates.

Large, long-term experiments are the best way to test causality, because they can alter just one variable (such as weight) while holding constant other factors that could confound the results. Obesity researchers have conducted few of these so-called randomized, controlled trials. "We don't know what happens when you turn fat people into thin people," Campos says. "That is not some oversight; there is no known way to do it"—except surgeries that carry serious risks and side effects.

"About 75 percent of American adults are trying to lose or maintain weight at any given time," reports Ali H. Mokdad, chief of the CDC's behavioral surveillance branch. A report in February by Marketdata Enterprises estimated that in 2004, seventy-one million Americans were actively dieting and that the nation spent about $46 billion on weight-loss products and services.

Dieting has been rampant for many years, and bariatric surgeries have soared in number from 36,700 in 2000 to roughly 140,000 in 2004, according to Marketdata. Yet when Flegal and others examined the CDC's most recent follow-up survey in search of obese senior citizens who had dropped into a lower weight category, they found that just 6 percent of nonobese, older adults had been obese a decade earlier.

Campos argues that for many people, dieting is not merely ineffective but downright counterproductive. A large study of nurses by Harvard Medical School doctors reported last year that 39 percent of the women had dropped weight only to regain it; those women later grew to be ten pounds heavier on average than women who did not lose weight.

Weight-loss advocates point to two trials that in 2001 showed a 58 percent reduction in the incidence of type 2 diabetes among people at high risk who ate better and exercised more. Participants lost little weight: an average of 2.7 kilograms after two years in one trial, 5.6 kilograms after three years in the other.

"People often say that these trials proved that weight loss prevents diabetes. They did no such thing," comments Steven N. Blair, an obesity researcher who heads the Cooper Institute in Dallas. Because the trials had no comparison group that simply ate a balanced diet and exercised without losing weight, they cannot rule out the possibility that the small drop in subjects' weights was simply a side effect. Indeed, one of the trial groups published a follow-up study in January that concluded that "at least 2.5 hours per week of walking for exercise during follow-up seemed to decrease the risk of diabetes by 63 to 69 percent, largely independent of dietary factors and BMI."

"H. L. Mencken once said that for every complex problem there is a simple solution—and it's wrong," Blair muses. "We have got to stop shouting from the rooftops that obesity is bad for you and that fat people are evil and weak-willed and that the world would be lovely if we all lost weight. We need to take a much more comprehensive view. But I don't see much evidence that that is happening."

MICHAEL SPECTER

Nature's Bioterrorist

FROM *THE NEW YORKER*

If as feared the avian flu becomes a world-wide epidemic, then any hope for neutralizing its spread and lethality will lie in the work done by doctors and researchers who have been studying it for the last several years. Michael Specter meets the men and women on the front lines of this emerging biological threat.

Early last September, an eleven-year-old girl from Kamphaeng Phet, a remote village in Thailand, developed a high fever, a severe cough, and a sore throat. She lived with her aunt and uncle in a one-room wooden house—not much more than a hut on stilts. The family had fifteen chickens, which wandered freely beneath the plank floor, where the young girl often played and slept. Then, at the end of August, the chickens died. Within days, the girl was sick, too. Her aunt took her to the hospital, but the fever kept rising. The girl's mother, who lived near Bangkok, where she worked at a factory, rushed to her bedside; sixteen hours later, her daughter was dead. In keeping with Thai custom, she was cremated immediately.

Avian influenza is nothing new in Thailand, or anywhere else where poultry are raised. Veterinarians often refer to it as the fowl plague, because in one form or another the disease has killed millions of chickens, turkeys, and other birds over the years. In 1983, the virus raced so rapidly through the Pennsylvania poultry population that health officials there were forced to slaughter nearly every chicken in the state. Until recently, however, humans rarely became infected with this type of virus. It had happened fewer than a dozen times since 1959, and in each case the illness was mild. But the strain that killed the girl from Kamphaeng Phet is different; in the past two years, it has caused the deaths of hundreds of millions of animals in nearly a dozen Asian countries. No such virus has ever spread so quickly over such a wide geographical area. Most viruses stick to a single species. This one has already affected a more diverse group than any other type of flu, and it has killed many animals previously thought to be resistant: blue pheasants, black swans, turtledoves, clouded leopards, mice, pigs, domestic cats, and tigers. Early in February, nearly five hundred open-billed storks were found dead in Thailand's largest fresh-water swamp, the Boraphet Reservoir. And the disease is no longer limited to Asia. In October, customs officers at the Brussels airport seized two infected eagles that had been smuggled from Thailand and destroyed them, along with the other animals held in quarantine at the airport.

This virus also kills people—so far, forty-two have died, including thirteen Vietnamese since Christmas. Those deaths represent more than three-quarters of all known avian-flu infections—an ominous mortality rate. Strains of influenza are named for two proteins on their surface that latch onto respiratory cells and permit the virus to invade them, and, if this strain, known as H5N1, becomes capable of spreading efficiently among humans, it will kill millions. Public-health officials in Asia, still reeling from the crisis created in 2003 by SARS (severe acute respiratory syndrome), have struggled to contain the burgeoning epidemic. Yet the task is immense. SARS was caused by a virus that turned out to be less deadly, less contagious, and far less aggressive than the flu. The threat was never as great. There are six billion chickens in Southeast Asia, and millions of households depend upon them, for income as well as for food; preventing this flu virus from spreading has become all but impossible. By early February, during Tet, Vietnam's biggest holiday, officials began posting livestock inspectors on major highways. Last week, officials banned poultry-raising in Ho Chi Minh City.

Vigilance is one of the few weapons available. Two weeks after the girl died in Kamphaeng Phet, Thai epidemiologists were asked to visit a hospital near Bangkok, where a woman had symptoms that matched those caused by the virus. It turned out to be a false alarm, but while the investigators were there a nurse took one of the doctors aside and mentioned that another woman had just died of similar flu symptoms. The death hadn't been reported, but the victim's last name sounded familiar, and so did the name of her village.

"It was just a fluke," Scott Dowell told me not long ago over tea in his office, on the sprawling campus of the Thai Ministry of Public Health, in the suburbs north of Bangkok. "Sure enough, the woman was the mother of that eleven-year-old girl. We would never have known if that nurse didn't happen to mention it." Dowell is the director of the Thailand office of the International Emerging Infections Program, which was established by the Centers for Disease Control in 2001. The group, which was among the first to identify SARS when that new virus appeared, is a front-line outpost dedicated to preventing the spread of infectious disease in a world where modern travel, global commerce, and porous borders mean that any new pathogen—whether a lethal strain of influenza, HIV, or the synthetic creation of a bioterrorist—may be no more than a plane ride away. Thailand is not nearly as poor as some of its neighbors, but its fertile climate and lengthy borders with Cambodia and Laos—neither of which has a credible public-health system—produce a continual stream of odd, and often deadly, infectious diseases. The Nipah virus, which is named for the village in Malaysia where it was discovered, in 1999, has been endemic; so have dengue and leptospirosis, a bacterial infection that affects humans and animals. Thailand also has serious epidemics of tuberculosis and HIV.

Now there is little on people's minds but the flu. "Our Thai colleagues called us and said, 'Hey, do you want to go to Kamphaeng Phet?' " Dowell said. "And on the way up there they told me the story. And I just said, 'We absolutely have to get hold of this virus.' This could be the first clear case of person-to-person transmission, the beginning of something very significant, something terrifying. So the girl had died and been cremated, and the mother was dead and embalmed. She was literally at the wat, awaiting cremation. The Thai doctor who was riding in the car with me was calling on his mobile phone to the team there, telling them to do whatever they had to do to hold that embalmed body out and take some blood or a tissue sample."

A PANDEMIC IS THE VIRAL EQUIVALENT of a perfect storm. There are three essential conditions, which rarely converge, and they are impossible to predict. But the requirements are clear. A new flu virus must emerge from the animal reservoirs that have always produced and harbored such viruses—one that has never infected human beings and therefore one to which no person would have antibodies. Second, the virus has to actually make humans sick. (Most don't.) Finally, it must be able to spread efficiently—through coughing, sneezing, or a handshake. For H5N1, the first two conditions have been met; it's new and it's deadly. On the ride to Kamphaeng Phet, Dowell couldn't stop wondering whether the virus had met the third condition, too.

If so, there would be little time to distribute medicine, develop a vaccine, establish quarantines, and plan to care for the millions—maybe tens of millions—of sick and dying people throughout the world. "That was why it was so important to get blood or tissue samples, for us to be able to know how this woman died," Dowell said. "Then we get up to the province and we go to the hospital where the girl was and the mother is dead and in comes the aunt and she has had a cough and fever for five days and she is complaining of a sore throat." Dowell and his colleagues grew even more alarmed. What had at first seemed like a coincidence now looked more like the start of an epidemic. "It really had to be human transmission, because there were no other cases in the village," Dowell told me. "Nobody tested positive for the virus. All the chickens had been killed, and the mother didn't even live there; she never encountered a chicken. There was just no other way for her to get sick."

Infectious-disease experts talk about pandemics the way geologists talk about earthquakes; the discussion is never about *whether* "the big one" will hit. Pandemics have recurred in cycles for centuries. The great flu of 1918 killed at least fifty million people and infected hundreds of millions of others. In the United States, it took more lives than all combat in the twentieth century. Hospitals and medical staffs were overwhelmed. Treatment of other diseases stopped. Less severe pandemics occurred in 1957 and in 1968, but each killed millions. The evidence in Thailand seemed to point toward a single conclusion. As it turned out, however, the virus was contained. The third condition for a pandemic had not been met: the aunt, who eventually recovered, and the mother, who did not, almost certainly became sick because each had spent so many hours with the girl; the virus had not mutated in such a way that it could pass easily among people.

Now, speaking in his office, Dowell was calm. "The world just has no idea what it's going to see if this thing comes," he said. Then he stopped himself and started over. "When, really. It's when. I don't think we can afford the luxury of the word 'if' anymore. We are past ifs. Whether it's tomorrow or next year or some other time, nobody knows for sure. The clock is ticking. We just don't know what time it is."

ON OCTOBER 14 the 441 black-striped Bengal tigers at the Sriracha Tiger Zoo, near Bangkok, began to die. They had been fed raw chicken carcasses from a local slaughterhouse. Each day, four or five developed symptoms of severe influenza. By the end of the month, forty-five were dead; more than a hundred others were infected and considered a danger to the remaining tigers as well as to the humans who tend them. They, too, had to be destroyed. Until 2003, tigers had never been susceptible to avian influenza. H5N1 was evolving rapidly and with unsettling implications.

I had stopped at the offices of the World Health Organization, in Geneva, on my way to Thailand. The scientists there were holding a meeting to discuss the feasibility of developing a vaccine that would lessen the impact of a pandemic. The man in charge of the WHO influenza program, Klaus Stöhr, was worried. Despite the mounting signs of danger, few countries or companies had taken the possibility of a pandemic seriously, and there was little interest in developing a vaccine. Stöhr has one of the trickiest jobs in the world of public health: remaining unruffled while racing to prepare for the worst. It is never easy to sound the alarm for an epidemic that might or might not soon sweep the earth. When I asked if Stöhr, who trained as a veterinarian, could assess the danger that H5N1 posed for humans, he sat down and sighed. "Let's take a step back for a moment," he said. "We know that during the last three pandemics—1918, 1957, 1968—all these viruses had genes that derived from animals. In 1918, the virus had consisted almost wholly of avian genes. Now we have a very powerful animal virus circulating in Asia that is already moving into humans. It is very widespread. The virus has changed its characteristics over the past few months. It is now highly pathogenic to chickens. It is infecting an increasing number of species—we now know that cats are susceptible. And they don't just start coughing and sneezing, either. They die."

He went on, "We know that the virus may have got a foothold in ducks, many of which, unlike chickens, appear to stay healthy. That's good news for ducks, but it can be very bad news for us, because controlling a disease that is spread through a hidden reservoir of highly mobile birds is much more difficult." Ducks are particularly dangerous, because they can carry in their digestive systems each strain of the virus—there are fifteen—that affects humans. Stöhr compared them to Trojan horses deployed throughout the world. I asked whether any of this information suggested that the virus was on the verge of a global assault on humanity. "Well, we can't say with certainty," he replied. "But take the example of a football player. Maybe he is even a bad football player. So he's kicking the ball ten, fifteen, twenty times toward the goal before he scores. And a good player does it once. This virus is not a good scorer. But even the worst scorer is going to make it one time. And in this game it's going to take only one goal for the virus to win."

IT IS HARD TO OVERSTATE THE DAMAGE that the death of sixty million chickens has caused to Thailand's national psyche. Until last year, the country had been the world's fourth-largest exporter of poultry. That ended with the first reliable reports of infection. Both the European Union and the United States have suspended imports from several countries of Southeast Asia. (China, which may have an even more serious problem, no longer exports poultry.) In Thailand, although there are many commercial producers, a large part of the poultry population remains with families and in small households. The Food and Agriculture Organization of the United Nations has estimated that two hundred million farmers in the region keep an average of about fifteen birds each—ducks, chickens, geese, turkeys, and quail. Most of these birds are free to scavenge in their yards, making them prone to infection from the migrating fowl that travel the seasonal flyways from Siberia and northern China.

There is a natural tension between the people who are responsible for a country's commercial agricultural production and those whose job is to safeguard public health. Chinese leaders, worried about trade and tourism, lied about SARS for months when it first appeared—ensuring that the virus would spread. Thailand has just put more than a hundred million dollars into the fight against avian influenza, but when early reports of an illness characteristic of the virus first surfaced there, more than a year ago, that government, too, was reluctant to act.

"It was difficult to persuade officials in this country to take the problem seriously," Prasert Thongcharoen, one of Thailand's most distinguished microbiologists, told me shortly after I'd arrived in Bangkok. We met one morning in his spartan office at Siriraj Hospital, on the Chao Phraya River. "I believe the Department of Livestock at first covered it up. I talked to farmers. At the very first sign of a problem, the department told the public that the chickens had cholera. But farmers said it wasn't cholera. If a chicken has cholera, you give it antibiotics and it gets better. These birds got sick one evening and the next day they were dead. That's not cholera. So I believe it was known, and it spread here at least from October 2003. One year ago or more. And why didn't the veterinarians realize that? It's not difficult to make that diagnosis. They didn't do the right thing. I'm not saying it would have stopped an epidemic, but they didn't do what they should have done."

THE NEXT MORNING I went with a woman from the World Health Organization to visit some farmers in Suphan Buri, a province about sixty-five miles north of Bangkok—a straight shot on the wide-open, recently completed freeway. Water buffalo grazed by roadside thickets of bamboo. Acres of bright-green rice paddies bent low in the breeze. Suphan Buri is a relatively prosperous province, thanks to the efforts of Banharn Silpa-archa, the former prime minister, who is from the area and whose political party long controlled agricultural development. It is a center of the Thai poultry industry and has been devastated by the epidemic. In Suphan Buri, chickens and ducks had been living in groups of between five thousand and ten thousand, often in long wooden coops spread out on stilts, and perched on top of a pond.

Finding a chicken in Suphan Buri proved to be a challenge. Farmers had been forced out of the business and were unwilling to talk. Most of their flocks had been killed. The government compensated the farmers generously, but the money didn't make up for their losses or persuade many to switch to a more regimented way of raising birds: coops now have to be built on land, where chicken feces can't contaminate the water. It's hard to change the habits of a nation, particularly when it costs a great deal and may be futile. After driving for about an hour—from empty farm to shuttered coop—we came upon a seventy-year-old farmer working a fishpond next to an abandoned chicken house. The man had deep-brown skin and a remarkably youthful face. He wore a long-sleeved white shirt and black knee-high rubber boots over clay-colored cotton pants. He had a dark-blue baseball cap on his head. Like many of his neighbors, he was a contract farmer who raised chickens for one of the conglomerates. Influenza had ruined him, he told me; all forty thousand of his chickens had died. He hired an earth mover to stuff the chickens into their coffins—giant blue-and-orange feed bags. "I could put thirty chickens in each bag," he said. He then used a crane to lift the bags into mass graves next to a copse of bamboo on his property. He pointed to a low rise and shook his head. "It's damn hard to watch," he said. "One day, they're all alive and healthy—the vets were here the week before to check them—and the next day they're dying by the thousand. It happened so quickly. They started shivering, thousands of them at once. And then they started to fall. Every one of them. They just fell over, dead." He said that he's stopped raising chickens, and that he's fed up with the government. He can't afford to switch to the enclosed system of housing poultry that Thailand now requires. "I'm too old, and you have to inspect everything every day," he said. "It's ridiculous. Under the new rules, you have to change clothes even when you go into one of the chicken houses. Who do you think is going to do that?"

I asked who he thought was responsible for the current situation. "Well, the bird flu came here from Hong Kong, like most diseases do," he said. "And SARS too. But once it's here I'm not sure how you get rid of it. And I'm not sure how you keep it from coming. They have killed millions of chickens, but are they going to kill every single one?"

The farmer is not entirely out of the business; he still raises birds for cockfighting, which is a big sport in Thailand. It is also a dangerous one when an avian virus is loose. These animals are often sold in the market, their cages rarely disinfected. People move from farm to farm to buy them. And now that such activity has become harder, middlemen are often paid to do it. Thousands of people attend cockfights. The government has tried to crack down, citing the ease with which the birds pass viruses to one another and to other animals in the marketplace. But it has not been very successful.

Thc man shrugged. He is now resigned to raising fish. "Chickens you had for forty days and then sold," he said. "Fish we have to raise for five months." He has sixty thousand fish in his pond. It's hard to know how much food to give them and how to keep birds from swooping down at feeding time and stealing it. "This whole thing is not for me," he said. "I want it back the way it was."

After leaving his farm, we were turned away at several others. The longhouses in the paddies all seemed empty. Eventually, for fifty baht—a little more than a dollar—I persuaded a man on a motorcycle to lead us toward the tree line, across a small bridge, and into a deep-green pasture next to a brook. The field was filled with some two thousand dark-brown long-billed ducks, penned in by a white picket fence. The din was enormous, even with the car windows closed. The man watching over the birds was protective of the owner, but he finally told us how to find her.

Saijai Phetsringharn is a forty-year-old woman who wears diamond rings on her fingers and gold bracelets on her wrists. As a wholesaler of duck eggs—she earns about two baht apiece—Phetsringharn is something of an avian activist, and she has strong opinions about how to stop the epidemic, which she has voiced frequently in public protests. "Vaccinate the animals," she said. "The government needs to vaccinate the chickens."

Vaccines for chickens do exist, but there are no consistent guidelines on how to use them. Both China and Indonesia—which do not export poultry—vaccinate their birds. But, so far, the Thai government has refused to adopt the practice. There are legitimate reasons for that: it takes time for the vaccines to work, and they may not always prevent the birds from becoming infected or from passing the virus on. In theory, that means humans could be exposed to healthy-looking birds that nonetheless carry the virus. Still, many public-health officials, including Scott Dowell, of the CDC, told me that it was time for the Thai government to reconsider the policy. Some villages are taking matters into their own hands, using herbs on the chickens, which many think will protect them.

"The government seems to think they can solve this problem just by killing birds," Phetsringharn said, mentioning that eight thousand of her ducks had recently been killed. "They are wrong. If they do that, the virus will persist but our industry will die."

She may be right, but recent genetic studies of the virus have determined that, while the ducks remain healthy, they are now shedding larger amounts of the virus and for longer periods than they did just a year ago. Last week, the government proposed a vast new culling—of three million free-range ducks. "Ducks are not like chickens," Phetsringharn said. "They need their freedom." She was standing in a garage stacked high with trays of large brown eggs. Many boxes of fowl-cholera vaccine were lined up on the floor. She is at least the third generation of her family to raise ducks. "We did it before I was born. Before my mother was born." She has three children, who are four, ten, and thirteen, and she expects them to enter the business, too. "If there is a business," she said, frowning.

INFLUENZA IS THE SIXTH-LEADING CAUSE of death in the United States, and it is responsible for even more damage in less-developed countries. During most flu seasons, as much as 20 percent of the American population becomes infected, about thirty-six thousand people might die, and more than two hundred thousand are admitted to hospitals. Few viruses have endured as long or done as much damage. Records of respiratory ailments that appear to have been influenza go back to ancient Greece and Rome. The word "influenza" derives from the Latin *influentia*—"influence"—reflecting the common belief that the epidemics it caused were due to the influence of the stars. People—often even doctors—refer casually to any respiratory infection as "the flu," but most are not. Influenza is caused by an orthomyxovirus, which comes in three types, designated A, B, and C. The B and C forms can infect people and make them sick, but they're not common and they're rarely serious. Type A is the virus we worry about. Every influenza virus has hundreds of microscopic spikes rising from its surface. Most are made of a viral protein called hemagglutinin, which can latch onto cells that the virus seeks to enter. The other spikes are called neuraminidase, an enzyme that helps the virus spread. These two proteins are the reason that flu viruses are labeled with the letters "H" and "N." Type A influenza has been so successful for so long because it is among the most mutable of viruses, capable of swapping or altering one or more of its eight genes with those from other strains.

Because this virus evolves so quickly, an annual flu shot is at best a highly educated bet on which strain is most likely to infect you. The vaccine stimulates antibodies that should provide protection from the particular strain of the virus that epidemiologists think will predominate this year. But if you are infected with a flu virus whose surface proteins have changed, your antibodies won't recognize them fully. That new strain could edge its way past the human immune system's complicated defenses and establish a new infection, and—though you might have some resistance, depending on how the strain had changed—you would need an entirely new set of antibodies to fight it. This goes on throughout our lives, and these small changes on the surface of the virus—the antigen—are called "antigenic drift."

The eight viral flu genes are put together in segments a bit like a line of connected Lego blocks, and they are easily dismantled, changed, and reassembled. When animal strains of influenza mix with human strains, there is always the possibility that the result will be an entirely new virus. That is called "antigenic shift." When large fragments of genetic material are replaced with genes from other influenza subtypes, or with genes from other animals, like pigs or chickens, the outcome is something that the human immune system will be unable to recognize. And even with the sophisticated tools of molecular genetics, we cannot predict how a virus will change or when or whether it will become more or less dangerous. We don't even know if survival of the fittest, when it comes to viruses, means survival of the most virulent: a virus so powerful that it kills all its hosts couldn't last long.

The elaborate transportation system for a flu virus adds to the chance nature of its evolution. The midpoint between Thailand and the Northern Hemisphere is Guangdong province, in southern China. Migrating birds often stop there, which may be one reason that so many epidemics begin in the region. The chances that an animal virus will jump the species barrier are greater in places with constant contact between humans and animals. Sixty percent of the world's people—and billions of its animals—live in Asia. There are eighty-six million people in Guangdong alone—more than the population of Germany. Every day, tens of thousands of chickens are moved by truck from Guangdong to Hong Kong. Agriculture, too, plays a role. Rice, the ubiquitous crop of Asia, needs water. Farmers raise ducks on the rice farms and on their ponds, and wild waterfowl can transmit their virus to domestic ducks there. The ducks can then pass it along to chickens and pigs, which often serve as a mixing vessel for human- and animal-flu viruses, because the receptors on their respiratory cells are similar to ours. In such an environment, one sneeze from a pig could be enough to start a pandemic.

WHEN THE BUSH ADMINISTRATION'S Health and Human Services Secretary Tommy Thompson announced his resignation, in December, he cited a potential epidemic of avian influenza as one of the greatest dangers facing the United States. The World Health Organization has put a "conservative estimate" of the deaths from such an event at between two and seven million people and expects that as many as a billion people would fall ill. Other numbers are even more sobering: Michael Osterholm, the director of the Center for Infectious Disease Research and Policy at the University of Minnesota, has made calculations based on the 1918 epidemic, taking into consideration general improvements in health care. Applying those fatality rates to the world's current population, Osterholm suggests that at least a hundred and eighty million people would die in a pandemic of similar severity. Shigeru Omi, the WHO official in charge of Asia, recently arrived at similar conclusions, stating that an H5N1 pandemic could infect between 25 percent and 30 percent of the world's population. Economic losses would be in the tens of billions of dollars.

Robert Webster, who holds a chair in virology at St. Jude Children's Research Hospital in Memphis, has been studying avian influenza for decades. Nobody knows more about the origins of the virus, how it changes, or what those changes might mean. "This is the worst flu virus I have ever seen or worked with or read about," he told me when we spoke recently. "We have to prepare as if we were going to war—and the public needs to understand that clearly. This virus is playing its role as a natural bioterrorist. The politicians are going to say Chicken Little is at it again. And, if I'm wrong, then thank God. But if it does happen, and I fully expect that it will, there will be no place for any of us to hide. Not in the United States or in Europe or in a bunker somewhere. The virus is a very promiscuous and efficient killer. That much we have known since 1997."

Pandemics seem to occur every thirty or forty years. On May 21, 1997, three decades after the previous outbreak, a three-year-old boy in Hong Kong died from what doctors were fairly certain was the flu. The most densely packed city on Earth, Hong Kong has often served as a cauldron for viral diseases and a transit point for epidemics. In 1894, a plague killed more than a hundred thousand people there after arriving from China. The two flu pandemics that swept the world in the past fifty years each appeared first in China, then in Hong Kong. In 2003 SARS originated in China, incubated in Hong Kong, and then fanned out across the globe. Health authorities in Hong Kong have every reason to be vigilant about unusual signs of illness.

The boy's doctor had forwarded a routine sample of his respiratory secretions to researchers at the Department of Health. Something about the structure and arrangement of the viral strain seemed strange to them. Clearly, it was a flu virus, but not a strain the Hong Kong scientists had seen before. They sent samples to research centers in England and the Netherlands and to the CDC in Atlanta. It took three months before specialists at the three institutions could agree on what they were looking at through their microscopes. It was H5N1, which until then had affected only poultry. Not even farmers who handled sick birds every day had become ill.

As soon as the strain was identified, Keiji Fukuda, the chief flu epidemiologist at the CDC, left for Hong Kong. The possibilities frightened him, but he was also excited by the opportunity to investigate a genuine medical mystery. How could that virus have infected a three-year-old child? Fukuda and his colleagues from the CDC spent a month in Hong Kong, turning over every piece of evidence they could find. But they saw no other indication that the virus had moved into the human population. "It had not spread to even one other person, so we just wrote it off as a freak occurrence," Fukuda told me when I went to see him in Atlanta. Fukuda is wiry, quiet, and intense—characteristics that underscore his deadpan presentation of facts. "It was weird and a little scary," he told me. "But it was so unexpected to find it in a human being that we eventually decided it must have been one instance involving that boy and that would be the end of it."

The team returned to the United States and moved on to other work. Then, later that year, a fifty-four-year-old man was hospitalized in Hong Kong with the bird flu. By December 5 he was dead. Fukuda returned to Hong Kong with the same troubling question that would eventually preoccupy his colleagues in Thailand. "I wondered if this virus was now acting like other strains—maybe its genes had mixed with human genes and it was moving freely among people. I remember sitting on the plane and thinking, If this is really happening, my

God, what are we in for?" By late December of 1997, eighteen people had become infected in Hong Kong, and six of them died—a remarkably high mortality rate. Researchers couldn't figure out how the virus was getting to humans. In the past, scientists had believed that such mutations required pigs to serve as a mixing vessel. These new cases suggested that H5N1 didn't need an intermediary host; it seemed that the virus had acquired the ability to move directly from birds to people. "I don't know why I remember this, but I do," Fukuda told me long after the events had taken place. "Christmas that year was on a Thursday," he said. "It was a very bad day. There were too many reports of people getting sick. The city just had this bad feeling to it. A heavy feeling."

Fukuda thought that only drastic measures would stop an epidemic, and he said so. The Hong Kong agriculture department, realizing that the virus was spreading rapidly within the poultry markets of the city, agreed. Many bird species were housed together in the markets, an ideal environment for genes from different viruses to mix and mutate. On December 29 city offficals began to kill every chicken in Hong Kong. One and a half million birds were culled, then disinfected and buried in landfills. "All these trucks driving up to a medium-sized poultry farm," Fukuda recalled. "Everyone is covered with protective gear—moon suits and all. They put the birds in bags and then used gas on them. Then they were deposited in the landfills." The decision was risky, but it seemed to work. No one else became ill, and many public-health experts believed that the slaughter in Hong Kong had prevented a horrifying epidemic. For several years, there were no other reports of the virus infecting a human being; some people even dared to hope that H5N1 had been vanquished. Fukuda was not so sure. He knew how risky—and potentially dangerous—such assumptions about infectious diseases can be. Then, in early 2003, Fukuda's fears were confirmed. A nine-year-old boy and his father were hospitalized in Hong Kong with the bird flu. The boy recovered, but on February 17 his father died.

That month, officials from the World Health Organization's rapid-alert team began hearing reports of a "strange contagious disease" that had attacked scores of people in Guangdong province, which borders Hong Kong. A month earlier, geese had suddenly started to die in several Hong Kong parks. Klaus Stöhr, of the WHO, happened to be attending a meeting in Beijing to help China develop a national flu-vaccine strategy. As Stöhr listened to each province report its general-health information, he became increasingly alarmed. Finally, an official from Guangdong stood up and spoke about an outbreak of what seemed like an unusually powerful flu. People were dying, and it was difficult to understand why. "I just put two and two together and it added up," Stöhr told me a few months later. "I thought this must be H5N1 coming back in precisely the way we had all feared. It was our worst nightmare—and the world's."

In the end, the illness that the Chinese were talking about turned out to be SARS, and scientists throughout the infectious-disease community, though concerned about the dangers of the new virus, took a deep breath. But the confusion was understandable. The early symptoms of SARS seemed a lot like those of the flu. And nobody expected this entirely new virus. Hong Kong went into economic, social, and psychological shock. Taxi-drivers disinfected their cabs (and advertised the fact on banners that flew from their windows) each time they dropped off a passenger. Subway ridership plummeted, and people who were brave enough to use the system wore white cotton masks—which, like antibacterial hand lotions, were selling briskly at every kiosk in the city. Restaurants and hotels were either shuttered or nearly empty. Headlines flashed constantly across every television set: "Outbreak," "In the Shadow of the Virus," and "Fear Across Asia." For the first time, the WHO issued a global health alert that advised travelers to avoid Hong Kong and Guangdong province unless they had no choice. Billions of dollars in revenues were lost, unemployment reached record levels, tourism vanished, and the epidemic pushed the city of seven million to the edge of its third recession in a decade.

Since SARS first appeared, it has killed fewer than a thousand people; during that time, a larger number drowned in bathtubs and swimming pools in the United States. The fears weren't altogether irrational, however. SARS was the first serious, easily transmitted disease to emerge in the twenty-first century, and the first in decades to move from animals to humans while also mutating into an illness that could be spread simply by breathing. More important, perhaps, coming not long after the anthrax scare of 2001, SARS reminded a world of something that it seemed to have forgotten—how easily it could be disabled by disease.

Early one morning, I drove to Sa Kaeo, a province about two hours east of Bangkok, on the Cambodian border. There, and in a province near Laos, Scott Dowell's CDC team has established a sort of radar defense system for microbes—the International Emerging Infections Program. The idea came about simply enough: the United States wants to protect itself from foreign threats, and in the 1990s—after confronting a series of new viral diseases—public-health officials argued persuasively that such unexpected infections posed a threat that the United States was not prepared to meet. Zoonotic diseases—animal illnesses that infect humans and make them sick—caused particular concern. Since the viruses often appear in densely populated Third World countries, the CDC set up its first operation in Thailand. A second office opened last year in Kenya. "The feeling was that maybe if we were out there with our ear to the ground, listening and watching, we might be able to catch something before it's too late," Mark Simmerman, who is with the CDC in Bangkok, told me as we drove. "So we decided to look very carefully at the residents of these two provinces."

The IEIP team has more than a million residents under surveillance for signs of pneumonia and other bacterial infections. (The theory is that if you look at lung X-rays and blood work closely enough to detect pneumonia, you are also likely to detect other significant problems.) Every morning, in every hospital, a surveillance officer reviews the previous day's admission logs. Whenever there is a questionable notation in the reports, doctors review it further. Technicians in the field are equipped to scan the X-rays and send them electronically to labs where more sophisticated analysis is possible. There have already been some successes: the program has helped raise the profile of influenza, which had traditionally been ignored by most Thais; it has made it easier for the provincial leadership to convince people that sick birds can be dangerous to humans. The team helped identify SARS, of course, and Dowell was one of the authors of the initial report characterizing the virus.

The CDC selected Thailand as the first country for the program in part because the government there was agreeable and committed. But there are other reasons. Health care is good. There are emergency rooms, antibiotics, and clinics. That is not true in neighboring Cambodia, however. Each day, thousands of Cambodian workers surge into Thailand—and often return home within forty-eight hours. The movement is never one-way; gambling is illegal in Thailand but not in Cambodia. Travel from Bangkok has increased markedly during the past three years. "This is sort of the classic situation we are facing in the world today," Nancy Cox told me before I left for Thailand. Cox is the chief of the influenza branch at the CDC, and she has been a strong advocate for establishing these listening posts, which are modeled on several that already exist in the United States. "We have Laos and Cambodia, which have fewer resources and much less of an infrastructure to deal with something like avian influenza or to do surveilance than they do in Thailand. And that really poses a perpetual danger, because they could be having outbreaks there that we simply are unable to know about. We are dealing with a difficult situation, and one that, unfortunately, looks like it will be with us for some time."

Our first stop was a village about forty kilometers from the border. It was a hot, airless day, and people weren't moving unless they had to. Seventy-five percent of the residents keep birds. Thailand has nine hundred thousand volunteers to watch over the health of its provincial villages. Each volunteer is responsible for a few households and tries to monitor the health of the people and even of the animals in his or her domain. A weekly bird census is taken—when I was there, there were 84,563 birds in the district. "Paying close attention this way may make the difference," Simmerman told me. "Maybe this will be the place where we will, in the end, see a cluster where the virus has begun to change. And, if so, we can hope to contain it far earlier than if we waited to see something like that in Bangkok or Hong Kong."

I spent some time with one of the health volunteers, a woman named Samorn Santhhape, who was wearing a yellow polo shirt and a floral-print skirt. We sat for a while on the stoop of a house and watched the chickens clucking and strutting about on the barren ground in front of us. There were about ten houses within sight, and at all of them the scene was the same. Each family owned about twenty chickens.

"These people are worried about this flu," she told me. "But they are more worried about losing their birds and the way they live. They are sometimes resentful of government announcements that tell them what to do." Animals have been kept in this way in Thailand for centuries, and many villagers feel that a basic right has come under attack. Yet any government that ignored such a severe threat to its animals and its people would be negligent. "What did we do wrong?" the woman at whose house we sat asked. She offered us tea, then rice and fruit. "We love our animals and treat them as well as we can." That seems to be true. Santhhape told me that there was a feeling of hopelessness. "When the chickens do get sick, they are dead before we even have time to notice," she said. "Sometimes it feels like we are trying to halt what we cannot even see."

IF THE AVIAN EPIDEMIC DOES MOVE WIDELY into human populations, as many scientists have predicted, it will mark the first time the world has been able to anticipate a pandemic. For thousands of years, people have rarely known the causes of their illnesses; they have certainly never been warned that an epidemic—whether smallpox, plague, cholera, or influenza—was imminent. Viral genetics has changed that. We can follow the evolution of a virus on a molecular level and gauge its power. Researchers at the CDC have just begun crucial experiments in specially protected laboratories where they will attempt to juggle the genetic components of the H5N1 virus. There are two ways to do that. First, the team will infect tissue with both the bird virus and a common human flu virus and see what grows. They will also use the tools of genetics. Molecular biology now allows scientists to break a virus down to its genes; the researchers will disassemble H5N1 and mix it in a variety of combinations with human flu viruses. Then they will test the results on animals.

There are other preparations under way too, of course, but they face significant financial, ethical, and political obstacles. The Bush Administration has put hundreds of millions of dollars into flu preparedness. Largely, that commitment reflects a world after 2001, in which it is not hard to convince leaders that if a natural virus can move quickly from Asia to America so could one that is produced by terrorists. Vaccines would provide the most effective protection at the lowest cost. Yet even vaccines won't solve a crisis. First, the world does not yet have the capacity to produce hundreds of millions of doses of an annual flu vaccine at the same time that it prepares for a pandemic strain like H5N1. In fact, this year the United States had trouble simply producing the annual vaccine. That became painfully apparent last October, when the government was forced to announce that a plant owned by one of its two principal manufacturers—Chiron Corporation—had been shut down by British licensing authorities. This delayed annual shots for millions of Americans. Currently, the world can produce three hundred million doses of flu vaccine a year—but there are more than six billion people, and in 1957 the pandemic crossed the globe in less than eight months. The ideal solution would be to develop vaccines that could offer protection from all influenza subtypes. Many public-health officials argue that such a vaccine should become a basic goal, like finding vaccines for AIDS, tuberculosis, and malaria. Without that, even if we could produce a billion doses of a new flu vaccine, which would we make? H5N1 might evolve so rapidly that a shot that would protect us today could prove useless six months from now. And who would get those vaccines? The most rational public-health approach would be to vaccinate those who would first be exposed—health-care workers and people in the region where an epidemic has struck. That is unlikely to happen. "Can you see the developed world shipping all its vaccine to Vietnam in a genuine crisis?" one senior U.S. public-health official told me. "Who are you kidding?"

The antiviral drug oseltamivir phosphate should protect against avian influenza. (It binds to one of the surface proteins and prevents the virus from multiplying.) The U.S. government has enough of the drug to treat 2.3 million people. It's made by one company, and a course of treatment costs more than sixty dollars. If the infections were spreading in Chicago or London, flooding the city with the drug might help. But in Thailand, Cambodia, or Vietnam? Not many people in those countries could come up with that kind of money—or find the drug even if they could. "This is where we can really have a remarkable effect," Bruce Gellin told me. Gellin is the director of the U.S. National Vaccine Program Office, and he spends most of his time these days trying to persuade governments and public officials to make the possibility of a pandemic a priority. "We need to realize that the stuff will be in short supply but that we can use it effectively," he said. "My analogy is a spark and a squirt gun. If you aim properly, you can get the spark and be done with it. If you miss, though, the fire is going to spread and there is nothing you can do to stop it."

What if the world did pay enough attention to what seems like an obvious threat? With the current knowledge about viral genetics, can we think about preventing the transmission of an epidemic of H5N1 from animals to people, or from Africa or Asia to Europe and America? Or could we contemplate achieving similar success with the Ebola virus or hemorrahagic fevers or Nipah or a new form of a hantavirus or some other exotic, and unidentified, disease? If you look at the ProMED daily report on the Internet—most infectious-disease experts read it every morning—the answer would appear to be no. The report, compiled by the International Society for Infectious Diseases, collects and forwards information about novel or important outbreaks of infectious diseases wherever they are found. And they are found everywhere: in the past several weeks, papaya ring-spot virus has appeared in southern India, Newcastle disease has returned to the chickens of Japan, and in January the Canadian Food Inspection Agency reported two cases of mad-cow disease. A woman with melioidosis, which is endemic in Southeast Asia, died on the southwest Indian Ocean island of Mauritius, the first known death there from the disease. There is no information about how the bacteria—often cited as a potential agent for biological terrorism—traveled five thousand miles.

"We are certainly better than we ever were at detecting viruses," Supamit Chunsuttiwat, who is one of the senior infectious-disease officials at the Thai Ministry of Public Health, told me. "But we are also much better at spreading them. A hundred years ago, the Nipah virus would have simply emerged and died out; instead, it was transmitted to pigs and amplified. With modern agriculture, the pigs are transported long distances to slaughter. And the virus goes with them." Few, if any, organisms have benefited from travel more than viruses have. For nearly fifty million years, the earth has been separated into distinct continents. Those continents were essentially walled off by geographic barriers. A virus could emerge, evolve, and move between species. But it could never pick up and move from, say, China to Montana. When smallpox emerged three thousand years ago, it would have killed the population around it and stayed where it was; but in the fifteenth and sixteenth centuries sailing vessels carried smallpox—and measles and yellow fever—between the Old World and the New World.

THESE KINDS OF HEALTH PROBLEMS were supposed to have been dispensed with by now; by the late 1960s, the Surgeon General had even announced that it was time to close the book on infectious disease. Polio, measles, and tuberculosis, major causes of death for centuries, no longer posed a serious threat in the United States. Good nutrition and living conditions helped; so did clean water. More important, immunization had taken many common diseases off the map, and the discovery of antibiotics had provided almost miracle cures for others. By the time the World Health Organization declared that smallpox had been eradicated, in 1979, cures had come to be expected.

Yet infectious diseases may present more of a danger than ever before. Late last week, for example, the WHO reported that scores of miners had died of pneumonic plague over the past two months in the Democratic Republic of Congo—the largest such outbreak in decades. And in 2003 alone, in addition to the SARS epidemic, monkeypox broke out for the first time in the Western Hemisphere (monkeypox causes symptoms that are similar to those of smallpox, but—at least for now—it is not nearly as contagious); West Nile disease continued its insidious spread across North America; and mad-cow disease appeared for the first time in the United States.

Other parts of the world, too, are experiencing the effects of infectious diseases that have never before been encountered. Rift Valley fever, an acute viral disease, suddenly infected hundreds of people in Yemen and Saudi Arabia in 2000—the first time it had been seen outside the African continent. Epidemiologists were amazed. "We never thought we would study viruses in Arabia," a senior researcher at the CDC told me. "There is nothing there. We always called it the empty corridor." However, there has been a lot of traffic in animals there, and in the case of Rift Valley fever, wind may have been enough to spread the virus. New viruses are not the only problem. Yellow fever, which crossed the world in the water barrels of ships, almost stopped the construction of the Panama Canal. The virus remains endemic in the tropics despite many attempts to eradicate it.

Travel, transportation, trade, pollution, and ecological disruption all play a role in ensuring the constant flow of disease from one part of the world to another. Last year's tsunami has pushed the world's public-health agencies to their limits. It's not clear how they could respond to a sudden global epidemic as well.

When I was in Sa Kaeo, I sat for a while in the district office with several provincial health officials. On the wall was a giant poster filled with avian-flu information. "Chickens used to live in our backyards—they didn't travel much. Now, throughout the world, farms have become factories," Charun Boonyarithikarn, the chief of social medicine for the district, told me. "Millions of chickens are shipped huge distances every day. We can't stop every chicken or duck or pig. And they offer millions of opportunities for pathogens to find a niche." He shook his head and smiled. "People around here fly to Hanoi and Phnom Penh and Paris. They visit China. They travel. Well, we have to realize this and accept it: take a plane ride to Paris and you may be taking an epidemic along with you."

GARDINER HARRIS AND
ANAHAD O'CONNOR

On Autism's Cause, It's Parents vs. Research

FROM THE *NEW YORK TIMES*

There are few areas where the gulf between science and popular belief seems wider than with the debate over the causes of autism. Several studies have failed to establish a link between autism and the mercury found in vaccinations, but many parents of autistic children refuse to accept these results. Gardiner Harris and Anahad O'Connor find frustration on both sides of this issue.

Kristen Ehresmann, a Minnesota Department of Health official, had just told a State Senate hearing that vaccines with microscopic amounts of mercury were safe. Libby Rupp, a mother of a three-year-old girl with autism, was incredulous.

"How did my daughter get so much mercury in her?" Ms. Rupp asked Ms. Ehresmann after her testimony.

"Fish?" Ms. Ehresmann suggested.

"She never eats it," Ms. Rupp answered.

"Do you drink tap water?"

"It's all filtered."

Ms. Ehresmann's colleague, Patricia Segal-Freemann, spoke up. "Well, do you breathe the air?" Ms. Segal-Freeman asked, with a resigned smile. Several parents looked angrily at Ms. Ehresmann and Ms. Segal-Freeman, who left.

Ms. Rupp remained, shaking with anger. "That anyone could defend mercury in vaccines," she said, "makes my blood boil."

Public health officials like Ms. Ehresmann, who herself has a son with autism, have been trying for years to convince parents like Ms. Rupp that there is no link between thimerosal—a mercury-containing preservative once used routinely in vaccines—and autism.

They have failed.

The Centers for Disease Control and Prevention, the Food and Drug Administration, the Institute of Medicine, the World Health Organization, and the American Academy of Pediatrics have all largely dismissed the notion that thimerosal causes or contributes to autism. Five major studies have found no link.

Yet despite all evidence to the contrary, the number of parents who blame thimerosal for their children's autism has only increased. And in recent months, these parents have used their numbers, their passion, and their organizing skills to become a potent national force. The issue has become one of the most fractious and divisive in pediatric medicine.

"This is like nothing I've ever seen before," Dr. Melinda Wharton, deputy director of the National Immunization Program, told a gathering of immunization officials in Washington in March. "It's an era where it appears that science isn't enough."

Parents have filed more than 4,800 lawsuits—200 from February to April alone—pushed for state and federal legislation banning thimerosal, and taken out full-page advertisements in major newspapers. They have also gained the support of politicians, including Senator Joseph I. Lieberman, Democrat of Connecticut, and Representatives Dan Burton, Republican of Indiana, and Dave Weldon,

Republican of Florida. And Robert F. Kennedy Jr. wrote an article in the June 16 issue of *Rolling Stone* magazine arguing that most studies of the issue are flawed and that public health officials are conspiring with drug makers to cover up the damage caused by thimerosal.

"We're not looking like a fringe group anymore," said Becky Lourey, a Minnesota state senator and a sponsor of a proposed thimerosal ban. Such a ban passed the New York State Legislature this week.

But scientists and public health officials say they are alarmed by the surge of attention to an idea without scientific merit. The antithimerosal campaign, they say, is causing some parents to stay away from vaccines, placing their children at risk for illnesses like measles and polio.

"It's really terrifying, the scientific illiteracy that supports these suspicions," said Dr. Marie McCormick, chairwoman of an Institute of Medicine panel that examined the controversy in February 2004.

Experts say they are also concerned about a raft of unproven, costly, and potentially harmful treatments—including strict diets, supplements, and a detoxifying technique called chelation—that are being sold for tens of thousands of dollars to desperate parents of autistic children as a cure for "mercury poisoning."

In one case, a doctor forced children to sit in a 160-degree sauna, swallow sixty to seventy supplements a day, and have so much blood drawn that one child passed out.

Hundreds of doctors list their names on a Web site endorsing chelation to treat autism, even though experts say that no evidence supports its use with that disorder. The treatment carries risks of liver and kidney damage, skin rashes, and nutritional deficiencies, they say.

In recent months, the fight over thimerosal has become even more bitter. In response to a barrage of threatening letters and phone calls, the Centers for Disease Control has increased security and instructed employees on safety issues, including how to respond if pies are thrown in their faces. One vaccine expert at the Centers wrote in an internal e-mail message that she felt safer working at a malaria field station in Kenya than she did at the agency's offices in Atlanta.

THIMEROSAL WAS FOR DECADES the favored preservative for use in vaccines. By weight, it is about 50 percent ethyl mercury, a form of mercury most scientists consider to be less toxic than methyl mercury, the type found in fish. The amount of ethyl mercury included in each childhood vaccine was once roughly equal to the amount of methyl mercury found in the average tuna sandwich.

In 1999 a Food and Drug Administration scientist added up all the mercury that American infants got with a full immunization schedule and concluded that the amount exceeded a government guideline. Some health authorities counseled no action, because there was no evidence that thimerosal at the doses given was harmful, and removing it might cause alarm. Others were not so certain that thimerosal was harmless.

In July 1999 the American Academy of Pediatrics and the Public Health Service released a joint statement urging vaccine makers to remove thimerosal as quickly as possible. By 2001 no vaccine routinely administered to children in the United States had more than half of a microgram of mercury—about what is found in an infant's daily supply of breast milk.

Despite the change, government agencies say that vaccines with thimerosal are just as safe as those without, and adult flu vaccines still contain the preservative.

But the 1999 advisory alarmed many parents whose children suffered from autism, a lifelong disorder marked by repetitive, sometimes self-destructive behaviors and an inability to form social relationships. In 10 to 25 percent of cases, autism seems to descend on young children seemingly overnight, sometime between their first and second birthdays.

Diagnoses of autism have risen sharply in recent years, from roughly 1 case for every 10,000 births in the 1980s to 1 in 166 births in 2003.

Most scientists believe that the illness is influenced strongly by genetics but that some unknown environmental factor may also play a role.

Dr. Tom Insel, director of the National Institute for Mental Health, said: "Is it cell phones? Ultrasound? Diet sodas? Every parent has a theory. At this point, we just don't know."

In 2000 a group of parents joined together to found SafeMinds, one of several organizations that argue that thimerosal is that environmental culprit. Their cause has been championed by politicians like Mr. Burton.

"My grandson received nine shots in one day, seven of which contained thimerosal, which is fifty percent mercury as you know, and he became autistic a short time later," he said in an interview.

In a series of House hearings held from 2000 through 2004, Mr. Burton called the leading experts who assert that vaccines cause autism to testify. They included a chemistry professor at the University of Kentucky who says that dental fillings cause or exacerbate autism and other diseases and a doctor from Baton Rouge, Louisiana, who says that God spoke to her through an eighty-seven-year-old priest and told her that vaccines caused autism.

Also testifying were Dr. Mark Geier and his son, David Geier, the experts whose work is most frequently cited by parents.

Dr. Geier has called the use of thimerosal in vaccines the world's "greatest catastrophe that's ever happened, regardless of cause."

He and his son live and work in a two-story house in suburban Maryland. Past the kitchen and down the stairs is a room with cast-off, unplugged laboratory equipment, wall-to-wall carpeting, and faux wood paneling that Dr. Geier calls "a world-class lab—every bit as good as anything at NIH."

Dr. Geier has been examining issues of vaccine safety since at least 1971, when he was a lab assistant at the National Institutes of Health, or NIH. His résumé lists scores of publications, many of which suggest that vaccines cause injury or disease.

He has also testified in more than ninety vaccine cases, he said, although a judge in a vaccine case in 2003 ruled that Dr. Geier was "a professional witness in areas for which he has no training, expertise, and experience."

In other cases, judges have called Dr. Geiers' testimony "intellectually dishonest," "not reliable," and "wholly unqualified."

The six published studies by Dr. Geier and David Geier on the relationship between autism and thimerosal are largely based on complaints sent to the Disease Control Centers by people who suspect that their children were harmed by vaccines.

In the first study, the Geiers compared the number of complaints associated with a thimerosal-containing vaccine, given from 1992 to 2000, with the complaints that resulted from a thimerosal-free version given from 1997 to 2000. The more thimerosal a child received, they concluded, the more likely an autism complaint was filed. Four other studies used similar methods and came to similar conclusions.

Dr. Geier said in an interview that the link between thimerosal and autism was clear.

Public health officials, he said, are "just trying to cover it up."

SCIENTISTS SAY THAT the Geiers' studies are tainted by faulty methodology.

"The problem with the Geiers' research is that they start with the answers and work backwards," said Dr. Steven Black, director of the Kaiser Permanente Vaccine Study Center in Oakland, California. "They are doing voodoo science."

Dr. Julie L. Gerberding, the director of the Disease Control Centers, said the agency was not withholding information about any potentially damaging effects of thimerosal.

"There's certainly not a conspiracy here," she said. "And we would never consider not acknowledging information or evidence that would have a bearing on children's health."

In 2003, spurred by parents' demands, the CDC asked the Institute of Medicine, an arm of the National Academy of Sciences and the nation's most prestigious medical advisory group, to review the evidence on thimerosal and autism.

In a report last year, a panel convened by the institute dismissed the Geiers' work as having such serious flaws that their studies were "uninterpretable." Some of the Geiers' mathematical formulas, the committee found, "provided no information," and the Geiers used basic scientific terms like "attributable risk" incorrectly.

In contrast, the committee found five studies that examined hundreds of thousands of health records of children in the United States, Britain, Denmark, and Sweden to be persuasive.

A study by the World Health Organization, for example, examined the health records of 109,863 children born in Britain from 1988 to 1997 and found that children who had received the most thimerosal in vaccines had the lowest incidence of developmental problems like autism.

Another study examined the records of 467,450 Danish children born from 1990 to 1996. It found that after 1992, when the country's only thimerosal-containing vaccine was replaced by one free of the preservative, autism rates rose rather than fell.

In one of the most comprehensive studies, a 2003 report by CDC, scientists examined the medical records of more than 125,000 children born in the United States from 1991 to 1999. It found no difference in autism rates among children exposed to various amounts of thimerosal.

Parent groups, led by SafeMinds, replied that documents obtained from the Disease Control Centers showed that early versions of the study had found a link between thimerosal and autism.

But CDC researchers said that it was not unusual for studies to evolve as more data and controls were added. The early versions of the study, they said, failed to control for factors like low birth weight, which increases the risk of developmental delays.

The Institute of Medicine said that it saw "nothing inherently troubling" with the CDC's adjustments and concluded that thimerosal did not cause autism. Further studies, the institute said, would not be "useful."

Since the report's release, scientists and health officials have been bombarded with hostile e-mail messages and phone calls. Dr. McCormick, the chairwoman of the institute's panel, said she had received threatening mail claiming that she was part of a conspiracy. Harvard University has increased security at her office, she said.

An e-mail message to the CDC on November 28 stated, "Forgiveness is between them and God. It is my job to arrange a meeting," according to records obtained by the *New York Times* after the filing of an open records request.

Another e-mail message, sent to the CDC on August 20, said, "I'd like to know how you people sleep straight in bed at night knowing all the lies you tell & the lives you know full well you destroy with the poisons you push & protect with your lies." Lyn Redwood of SafeMinds said that such e-mail messages did not represent her organization or other advocacy groups.

In response to the threats, CDC officials have contacted the Federal Bureau of Investigation and heightened security at the Disease Control Centers. Some officials said that the threats had led them to look for other jobs.

In *Evidence of Harm*, a book published earlier this year that is sympathetic to the notion that thimerosal causes autism, the author, David Kirby, wrote that the thimerosal theory would stand or fall within the next year or two.

Because autism is usually diagnosed sometime between a child's third and fourth birthdays and thimerosal was largely removed from childhood vaccines in 2001, the incidence of autism should fall this year, he said.

No such decline followed thimerosal's removal from vaccines during the 1990s in Denmark, Sweden, or Canada, researchers say.

But the debate over autism and vaccines is not likely to end soon.

"It doesn't seem to matter what the studies and the data show," said Ms. Ehresmann, the Minnesota immunization official. "And that's really scary for us because if science doesn't count, how do we make decisions? How do we communicate with parents?"

NEIL SWIDEY

What Makes People Gay?

FROM THE *BOSTON GLOBE MAGAZINE*

In the old debate over genes versus environment, there's a surprising new twist. Mounting evidence suggests that the key to sexual orientation may be the environment of the mother's womb. Neil Swidey surveys this shifting terrain of research.

WITH CRYSTAL-BLUE EYES, wavy hair, and freshly scrubbed faces, the boys look as though they stepped out of a Pottery Barn Kids catalog. They are seven-year-old twins. I'll call them Thomas and Patrick; their parents agreed to let me meet the boys as long as I didn't use their real names.

Spend five seconds with them, and there can be no doubt that they are identical twins—so identical even they can't tell each other apart in photographs. Spend five minutes with them, and their profound differences begin to emerge.

Patrick is social, thoughtful, attentive. He repeatedly addresses me by name. Thomas is physical, spontaneous, a bit distracted. Just minutes after meeting me outside a coffee shop, he punches me in the upper arm, yells, "Gray punch buggy!" and then points to a Volkswagen Beetle cruising past us. It's a hard punch. They horse around like typical brothers, but Patrick's punches are less forceful and his voice is higher. Thomas charges at his brother, arms flexed in front of him like a mini-bodybuilder. The differences are subtle—they're seven-year-old boys, after all—but they are there.

When the twins were two, Patrick found his mother's shoes. He liked wearing them. Thomas tried on his father's once but didn't see the point.

When they were three, Thomas blurted out that toy guns were his favorite things. Patrick piped up that his were the Barbie dolls he discovered at day care.

When the twins were five, Thomas announced he was going to be a monster for Halloween. Patrick said he was going to be a princess. Thomas said he couldn't do that, because other kids would laugh at him. Patrick seemed puzzled. "Then I'll be Batman," he said.

Their mother—intelligent, warm, and open-minded—found herself conflicted. She wanted Patrick—whose playmates have always been girls, never boys—to be himself, but she worried his feminine behavior would expose him to ridicule and pain. She decided to allow him free expression at home while setting some limits in public.

That worked until last year, when a school official called to say Patrick was making his classmates uncomfortable. He kept insisting that he was a girl.

Patrick exhibits behavior called childhood gender nonconformity, or CGN. This doesn't describe a boy who has a doll somewhere in his toy collection or tried on his sister's Snow White outfit once, but rather one who consistently exhibits a host of strongly feminine traits and interests while avoiding boy-typical behavior like rough-and-tumble play. There's been considerable research into this phenomenon, particularly in males, including a study that followed boys from an early age into early adulthood. The data suggest there is a very good chance Patrick will grow up to be homosexual. (Rather than transgender, as some might suspect.) Not all homosexual men show this extremely feminine behavior as young boys. But the research indicates that, of the boys who do exhibit CGN, about 75 percent of them—perhaps more—turn out to be gay or bisexual.

What makes the case of Patrick and Thomas so fascinating is that it calls into question both of the dominant theories in the long-running debate over what makes people gay: nature or nurture, genes or learned behavior. As identical twins, Patrick and Thomas began as genetic clones. From the moment they came out of their mother's womb, their environment was about as close to identical as possible—being fed, changed, and plopped into their car seats the same way, having similar relationships with the same nurturing father and mother. Yet before either boy could talk, one showed highly feminine traits while the other appeared to be "all boy," as the moms at the playgrounds say with apologetic shrugs.

"That my sons were different the second they were born, there is no question about it," says the twins' mother.

So what happened between their identical genetic starting point and their births? They spent nine months in utero. In the hunt for what causes people to be gay or straight, that's now the most interesting and potentially enlightening frontier.

WHAT DOES IT MATTER WHERE homosexuality comes from? Proving people are born gay would give them wider social acceptance and better protection against discrimination, many gay rights advocates argue. In the last decade, as this "biological" argument has gained momentum, polls find Americans—especially young adults—increasingly tolerant of gays and lesbians. And that's exactly what has groups opposed to homosexuality so concerned. The Family Research Council, a conservative Christian think tank in Washington, D.C., argues in its book *Getting It Straight* that finding people are born gay "would advance the idea that sexual orientation is an innate characteristic, like race; that homosexuals, like African Americans, should be legally protected against 'discrimination'; and that disapproval of homosexuality should be as socially stigmatized as racism. However, it is not true."

Some advocates of gay marriage argue that proving sexual orientation is inborn would make it easier to frame the debate as simply a matter of civil rights. That could be true, but then again, freedom of religion enjoyed federal protection long before inborn traits like race and sex.

For much of the twentieth century, the dominant thinking connected homosexuality to upbringing. Freud, for instance, speculated that overprotective mothers and distant fathers helped make boys gay. It took the American Psychiatric Association until 1973 to remove "homosexuality" from its manual of mental disorders.

Then, in 1991, a neuroscientist in San Diego named Simon LeVay told the world he had found a key difference between the brains of homosexual and heterosexual men he studied. LeVay showed that a tiny clump of neurons of the anterior hypothalamus—which is believed to control sexual behavior—was, on average, more than twice the size in heterosexual men as in homosexual men. LeVay's findings did not speak directly to the nature-vs.-nurture debate—the clumps could, theoretically, have changed size *because* of homosexual behavior. But that seemed unlikely, and the study ended up jump-starting the effort to prove a biological basis for homosexuality.

Later that same year, Boston University psychiatrist Richard Pillard and Northwestern University psychologist J. Michael Bailey announced the results of their study of male twins. They found that, in identical twins, if one twin was gay, the other had about a 50 percent chance of also being gay. For fraternal twins, the rate was about 20 percent. Because identical twins share their entire genetic makeup while fraternal twins share about half, genes were believed to explain the difference. Most reputable studies find the rate of homosexuality in the general population to be 2 to 4 percent, rather than the popular "1 in 10" estimate.

In 1993 came the biggest news: Dean Hamer's discovery of the "gay gene." In fact, Hamer, a Harvard-trained researcher at the National Cancer Institute, hadn't quite put it that boldly or imprecisely. He found that gay brothers shared a specific region of the X chromosome, called Xq28, at a higher rate than gay men shared with their straight brothers. Hamer and others suggested this finding would eventually transform our understanding of sexual orientation.

That hasn't happened yet. But the clear focus of sexual-orientation research has shifted to biological causes, and there hasn't been much science produced to support the old theories tying homosexuality to upbringing. Freud may have been seeing the effect rather than the cause, since a father faced with a very feminine son might well become more distant or hostile, leading the boy's mother to become more protective. In recent years, researchers who suspect that homosexuality is inborn—whether because of genetics or events happening in the womb—have looked everywhere for clues: Prenatal hormones. Birth order. Finger length. Fingerprints. Stress. Sweat. Eye blinks. Spatial relations. Hearing. Handedness. Even "gay" sheep.

LeVay, who is gay, says that when he published his study fourteen years ago, some gays and lesbians criticized him for doing research that might lead to homosexuality once again being lumped in with diseases and disorders. "If anything, the reverse has happened," says LeVay, who is now sixty-one and no longer active in the lab. He says the hunt for a biological basis for homosexuality, which involves many researchers who are themselves gay or lesbian, "has contributed to the status of gay people in society."

These studies have been small and underfunded, and the results have often been modest. Still, because there's been so much of this disparate research, "all sort of pointing in the same direction, makes it pretty clear there are biological processes significantly influencing sexual orientation," says LeVay. "But it's also kind of frustrating that it's still a bunch of hints, that nothing is really as crystal clear as you would like."

Just in the last few months, though, the hints have grown stronger.

In May, Swedish researchers reported finding important differences in how the brains of straight men and gay men responded to two compounds suspected of being pheromones—those scent-related chemicals that are key to sexual arousal in animals. The first compound came from women's urine, the second from male sweat. Brain scans showed that when straight men smelled the female urine compound, their hypothalamus lit up. That didn't happen with gay men. Instead, their hypothalamus lit up when they smelled the male-sweat compound, which was the same way straight women had responded. This research once again connecting the hypothalamus to sexual orientation comes on the heels of work with sheep. About 8 percent of domestic rams are exclusively interested in sex with other rams. Researchers found that a clump of neurons similar to the one LeVay identified in human brains was also smaller in gay rams than straight ones. (Again, it's conceivable that these differences could be showing effect rather than cause.)

In June, scientists in Vienna announced that they had isolated a master genetic switch for sexual orientation in the fruit fly. Once they flicked the switch, the genetically altered female flies rebuffed overtures from males and instead attempted to mate with other females, adopting the elaborate courting dance and mating songs that males use.

And now, a large-scale, five-year genetic study of gay brothers is underway in North America. The study received $2.5 million from the National Institutes of Health, which is unusual. Government funders tend to steer clear of sexual orientation research, aware that even small grants are apt to be met with outrage from conservative congressmen looking to make the most of their C-SPAN face time. Relying on a robust sample of 1,000 gay-brother pairs and the latest advancements in genetic screening, this study promises to bring some clarity to the murky area of what role genes may play in homosexuality.

This accumulating biological evidence, combined with the prospect of more on the horizon, is having an effect. Last month, the Rev. Rob Schenck, a prominent Washington, D.C., evangelical leader, told a large gathering of young evangelicals that he believes homosexuality is not a choice but rather a predisposition, something "deeply rooted" in people. Schenck told me that his conversion came about after he'd spoken extensively with genetic researchers and psychologists. He argues that evangelicals should continue to oppose homosexual behavior, but that "many evangelicals are living in a sort of state of denial about the advance of this conversation." His message: "If it's inevitable that this scientific evidence is coming, we have to be prepared with a loving response. If we don't have one, we won't have any credibility."

As the twenty-one-year-old college junior in a hospital johnny slides into the MRI, she is handed controls with buttons for "strongly like" and "strongly dislike." Hundreds of pornographic images—in male-male and female-female pairings—flash before her eyes. Eroticism eventually gives way to monotony, and it's hard to avoid looking for details to distinguish one image from the rest of the panting pack. So it goes from "Look at the size of those breasts!" to "That can't be comfortable, given the length of her fingernails!" to "Why is that guy wearing nothing but work boots on the beach?"

Regardless of which buttons the student presses, the MRI scans show her arousal level to each image, at its starting point in the brain.

Researchers at Northwestern University, outside Chicago, are doing this work as a follow-up to their studies of arousal using genital measurement tools. They found that while straight men were aroused by film clips of two women having sex, and gay men were aroused by clips of two men having sex, most of the men who identified themselves as bisexual showed gay arousal patterns. More surprising was just how different the story with women turned out to be. Most women, whether they identified as straight, lesbian, or bisexual, were significantly aroused by straight, gay, and lesbian sex. "I'm not suggesting that most women are bisexual," says Michael Bailey, the psychology professor whose lab conducted the studies. "I'm suggesting that whatever a woman's sexual arousal pattern is, it has little to do with her sexual orientation." That's fundamentally different from men. "In men, arousal is orientation. It's as simple as that. That's how gay men learn they are gay."

These studies mark a return to basics for the forty-seven-year-old Bailey. He says researchers need a far deeper understanding of what sexual orientation is before they can determine where it comes from.

Female sexual orientation is particularly foggy, he says, because there's been so little research done. As for male sexual orientation, he argues that there's now enough evidence to suggest it is "entirely inborn," though not nearly enough to establish how that happens.

Bailey's 1991 twin study is still cited by other researchers as one of the pillars in the genetic argument for homosexuality. But his follow-up study using a comprehensive registry of twins in Australia found a much lower rate of similarity in sexual orientation between identical twins, about 20 percent, down from 50 percent. Bailey still believes that genes make important contributions to sexual orientation. But, he says, "that's not where I'd bet the real breakthroughs will come."

His hunch is that further study of childhood gender nonconformity will pay big. Because it's unclear what percentage of homosexuals and lesbians showed CGN as children, Bailey and his colleagues are now running a study that uses adult participants' home movies from childhood to look for signs of gender-bending behavior.

Cornell psychologist Daryl Bem has proposed an intriguing theory for how CGN might lead to homosexuality. According to this pathway, which he calls "the exotic becomes erotic," children are born with traits for temperament, such as aggression and activity level, that predispose them to male-typical or female-typical activities. They seek out playmates with the same interests. So a boy whose traits lead him to hopscotch and away from rough play will feel different from, and ostracized by, other boys. This leads to physiological arousal of fear and anger in their presence, arousal that eventually is transformed from exotic to erotic.

Critics of homosexuality have used Bem's theory, which stresses environment over biology, to argue that sexual orientation is not inborn and not fixed. But Bem says this pathway is triggered by biological traits, and he doesn't really see how the outcome of homosexuality can be changed.

Bailey says whether or not Bem's theory holds up, the environment most worth focusing in on is the one a child experiences when he's in his mother's womb.

LET'S GET BACK TO THOMAS AND PATRICK. Because it's unclear why twin brothers with identical genetic starting points and similar post-birth environments would take such divergent paths, it's helpful to return to the beginning.

Males and females have a fundamental genetic difference—females have two X chromosomes, and males have an X and a Y. Still, right after conception, it's hard to tell male and female zygotes apart, except for that tucked-away chromosomal difference. Normally, the changes take shape at a key point of fetal development, when the male brain is masculinized by sex hormones. The female brain is the default. The brain will stay on the female path as long as it is protected from exposure to hormones. The hormonal theory of homosexuality holds that, just as exposure to circulating sex hormones determines whether a fetus will be male or female, such exposure must also influence sexual orientation.

The cases of children born with disorders of "sexual differentiation" offer insight. William Reiner, a psychiatrist and urologist with the University of Oklahoma, has evaluated more than a hundred of these cases. For decades, the standard medical response to boys born with severely inadequate penises (or none at all) was to castrate the boy and have his parents raise him as a girl. But Reiner has found that nurture—even when it involves surgery soon after birth—cannot trump nature. Of the boys with inadequate penises who were raised as girls, he says, "I haven't found one who is sexually attracted to males." The majority of them have transitioned back to being males and report being attracted to females.

During fetal development, sexual identity is set before the sexual organs are formed, Reiner says. Perhaps it's the same for sexual orientation. In his research, of all the babies with X and Y chromosomes who were raised as girls, the only ones he has found who report having female identities and being attracted to males are those who did not have "receptors" to let the male sex hormones do their masculinizing in the womb.

What does this all mean? "Exposure to male hormones in utero dramatically raises the chances of being sexually attracted to females," Reiner says. "We can infer that the absence of male hormone exposure may have something to do with attraction to males."

Michael Bailey says Reiner's findings represent a major breakthrough, showing that "whatever causes sexual orientation is strongly influenced by prenatal biology." Bailey and Reiner say the answer is probably not as simple as just exposure to sex hormones. After all, the exposure levels in some of the people Reiner studies are abnormal enough to produce huge differences in sexual organs. Yet sexual organs in straight and gay people are, on average, the same. More likely, hormones are interacting with other factors.

Canadian researchers have consistently documented a "big-brother effect," finding that the chances of a boy being gay increase with each additional older brother he has. (Birth order does not appear to play a role with lesbians.) So, a male with three older brothers is three times more likely to be gay than one with no older brothers, though there's still a better than 90 percent chance he will be straight. They argue that this results from a complex interaction involving hormones, antigens, and the mother's immune system.

By now, there is substantial evidence showing correlation—though not causation—between sexual orientation and traits that are set when a baby is in the womb. Take finger length. In general, men have shorter index fingers in relation to their ring fingers; in women, the lengths are generally about the same. Researchers have found that lesbians generally have ratios closer to males. Other studies have shown masculinized results for lesbians in inner-ear functions and eye-blink reactions to sudden loud noises, and feminized patterns for gay men on certain cognitive tasks like spatial perception and remembering the placement of objects.

New York University researcher Lynn S. Hall, who has studied traits determined in the womb, speculates that Patrick was somehow prenatally stressed, probably during the first trimester, when the brain is really developing, particularly the structures like the hypothalamus that influence sexual behavior. This stress might have been based on his position in the womb or the blood flow to him or any of a number of other factors not in his mother's control. Yet more evidence that identical twins have womb experiences far from identical can be found in their often differing birth weights. Patrick was born a pound lighter than Thomas.

Taken together, the research suggests that early on in the womb, as the fetus's brain develops in either the male or female direction, something fundamental to sexual orientation is happening. Nobody's sure what's causing it. But here's where genes may be involved, perhaps by regulating hormone exposure or by dictating the size of that key clump of neurons in the hypothalamus. Before researchers can sort that out, they'll need to return to the question of whether, in fact, there is a "gay gene."

THE CROWD ON BOSTON COMMON is thick on this scorcher of a Saturday afternoon in June, as the throngs make their way around the thirty-fifth annual Boston Pride festival, past booths peddling everything from "Gayopoly" board games to *Braveheart*ian garments called Utilikilts. Sitting quietly in his booth is Alan Sanders, a soft-spoken forty-one-year-old with a sandy beard and thinning hair. He's placed a mound of rainbow-colored Starbursts on the table in front of him and hung a banner that reads: "WANTED: Gay Men with Gay Brothers for Molecular Genetic Study of Sexual Orientation."

Sanders is a psychiatrist with the Evanston Northwestern Healthcare Research Institute who is leading the NIH-funded search for the genetic basis of male homosexuality (www.gaybros.com). He is spending the summer crisscrossing the country, going to gay pride festivals, hoping to recruit 1,000 pairs of gay brothers to participate. (His wife, who just delivered their third son, wasn't crazy about the timing.) When people in Boston ask him how much genes may contribute to homosexuality, he says the best estimate is about 40 percent.

Homosexuality runs in families—studies show that 8 to 12 percent of brothers of gay men are also gay, compared with the 2 to 4 percent of the general population.

Sanders spends much of the afternoon handing out Starbursts to people who clearly don't qualify for a gay brothers study—preteen girls, adult lesbians wearing T-shirts that read "I Like Girls Who Like Girls," and elderly women in straw hats who speak only Chinese. But many of the gay men who stop by are interested in more than free candy. Among the people signing up is James Daly, a thirty-one-year-old from Salem. "I think it's important for the public—especially the religious right—to know it's not a choice for some people," Daly says. "I feel I was born this way."

(In fairness, there aren't many leaders of groups representing social and religious conservatives who still argue that homosexual orientation—as opposed to behavior—is a matter of choice. Even as he insists that no one is born gay, Peter Sprigg, the point person on homosexuality for the Family Research Council, says, "I don't think that people choose their sexual attraction.")

In the decade since Dean Hamer made headlines, the gay gene theory has taken some hits. A Canadian team was unable to replicate his findings. Earlier this year, a team from Hamer's own lab reported only mixed results after having done the first scan of the entire human genome in the search for genes influencing sexual orientation.

But all of the gene studies so far have been based on small samples and lacked the funding to do things right. Sanders's study should be big enough to provide some real answers on linkage as well as shed light on gender nonconformity and the big-brother effect.

There is, however, a towering question that Sanders's study will probably not be able to answer. That has to do with evolution. If a prime motivation of all species is to pass genes on to future generations, and gay men are estimated to produce 80 percent fewer offspring than straight men, why would a gay gene not have been wiped out by the forces of natural selection? This evolutionary disadvantage is what led former Amherst College biologist Paul Ewald and his colleague Gregory Cochran to argue that homosexuality might be caused by a virus—a pathogen most likely working in utero. That argument caused a stir when they proposed it six years ago, but with no research done to test it, it remains just another theory. Other scientists have offered fascinating but unpersuasive explanations, most of them focusing on some kind of compensatory benefit, in the same way that the gene responsible for sickle cell anemia also protects against malaria. A study last year by researchers in Italy showed that female relatives of gay men tended to be more fertile, though, as critics point out, not nearly fertile enough to make up for the gay man's lack of offspring.

But there will be plenty of time for sorting out the evolutionary paradox once—and if—researchers are able to identify actual genes involved in sexual orientation. Getting to that point will likely require integrating multiple lines of promising research. That is exactly what's happening in Eric Vilain's lab at the University of California, Los Angeles. Vilain, an associate professor of human genetics, and his colleague, Sven Bocklandt, are using gay sheep, transgenic mice, identical twin humans, and novel approaches to human genetics to try to unlock the mystery of sexual orientation.

Instead of looking for a gay gene, they stress that they are looking for several genes that cause either attraction to men or attraction to women. Those same genes would work one way in heterosexual women and another way in homosexual men. The UCLA lab is examining how these genes might be turned "up" or "down." It's not a question of what genes you have, but rather which ones you use, says Bocklandt. "I have the genes in my body to make a vagina and carry a baby, but I don't use them, because I am a man." In studying the genes of gay sheep, for example, he's found some that are turned "way up" compared with the straight rams.

The lab is also testing an intriguing theory involving imprinted genes. Normally, we have two copies of every gene, one from each parent, and both copies work. They're identical, so it doesn't matter which copy comes from which parent. But with imprinted genes, that does matter. Although both copies are physically there, one copy—either from the mom or the dad—is blocked from working. Think of an airplane with an engine on each wing, except one of the engines is shut down. A recent Duke University study suggests humans have hundreds of imprinted genes, including one on the X chromosome that previous research has tied to sexual orientation.

With imprinted genes, there is no backup engine. So if there's something atypical in the copy from mom, the copy from dad cannot be turned on. The UCLA lab is now collecting DNA from identical twins in which one twin is straight and the other is gay. Because the twins begin as genetic clones, if a gene is imprinted in one twin, it will be in the other twin as well. Normally, as the fetuses are developing, each time a cell divides, the DNA separates and makes a copy of itself, replicating all kinds of genetic information. It's a complicated but incredibly accurate process. But the coding to keep the backup engine shut down on an imprinted gene is less accurate.

So how might imprinted genes help explain why one identical twin would be straight and the other gay? Say there's an imprinted gene for attraction to females, and there's something atypical in the copy the twin brothers get from mom. As all that replicating is going on, the imprinting (to keep the copy from dad shut down) proceeds as expected in one twin, and he ends up gay. But somehow with his brother, the coding for the imprinting is lost, and rather than remain shut down, the fuel flows to fire up the backup engine from dad. And that twin turns out to be straight.

IN THE COURSE OF REPORTING THIS STORY, I experienced a good deal of whiplash. Just when I would become swayed by the evidence supporting one discrete theory, I would stumble onto new evidence casting some doubt on it. Ultimately, I accepted this as unavoidable terrain in the hunt for the basis of sexual orientation. This is, after all, a research field built on underfunded, idiosyncratic studies that are met with full-barreled responses from opposing and well-funded advocacy groups determined to make the results from the lab hew to the scripts they've honed for the talk-show circuit.

You can't really blame the advocacy groups. The stakes are high. In the end, homosexuality remains such a divisive issue that only thoroughly tested research will get society to accept what science has to say about its origin. Critics of funding for sexual orientation research say that it isn't curing cancer, and they're right. But we devote a lot more dollars to studying other issues that aren't curing cancer and have less resonance in society.

Still, no matter how imperfect these studies are, when you put them all together and examine them closely, the message is clear: While postbirth development may well play a supporting role, the roots of homosexuality, at least in men, appear to be in place by the time a child is born. After spending years sifting through all the available data, British researchers Glenn Wilson and Qazi Rahman come to an even bolder conclusion in their book *Born Gay: The Psychobiology of Sex Orientation,* in which they write: "Sexual orientation is something we are born with and not 'acquired' from our social environment."

Meanwhile, the mother of twins Patrick and Thomas has done her own sifting and come to her own conclusions. She says her son's feminine behavior suggests he will grow up to be gay, and she has no problem with that. She just worries about what happens to him between now and then.

After that fateful call from Patrick's school, she says, "I knew I had to talk to my son, and I had no clue what to say." Ultimately, she told him that although he could play however he wanted at home, he couldn't tell his classmates he was a girl, because they'd think he was lying. And she told him that some older boys might be mean to him and even hit him if he continued to claim he was a girl.

Then she asked him, "Do you think that you can convince yourself that you are a boy?"

"Yes, Mom," he said. "It's going to be like when I was trying to learn to read, and then one day I opened the book and I could read."

His mother's heart sank. She could tell that he wanted more than anything to please her. "Basically, he was saying there must be a miracle—that one day I wake up and I'm a boy. That's the only way he could imagine it could happen."

In the year since that conversation, Patrick's behavior has become somewhat less feminine. His mother hopes it's just because his interests are evolving and not because he's suppressing them.

"I can now imagine him being completely straight, which I couldn't a year ago," she says. "I can imagine him being gay, which seems to be statistically most likely."

She says she's fine with either outcome, just as long as he's happy and free from harm. She takes heart in how much more accepting today's society is. "By the time my boys are twenty, the world will have changed even more."

By then, there might even be enough consensus for researchers to forget about finger lengths and fruit flies and gay sheep, and move on to a new mystery.

JONATHAN WEINER

The Tangle

FROM *THE NEW YORKER*

A disease that once afflicted people living on Guam has become one of the great unsolved mysteries in neurology. Now a researcher from outside the field thinks he has discovered the answer. Jonathan Weiner reports on this controversial theory and the maverick scientist behind it.

Billy Borja grew up in Sinajana, a hilltop village near the center of the island of Guam. As a teenager in the 1950s, he liked to hunt, and a few times a month he would take his shotgun down into the "boonies," which was what locals called the jungle that surrounded Sinajana. He almost always came back with a deer, a wild pig, or a dozen flying foxes—fruit bats with wingspans of three feet, which were a delicacy on Guam and on Rota, its nearest neighbor in the Mariana archipelago, and on many other islands in Oceania. Borja's mother boiled the bats in coconut milk and served them in the native Chamorro way: heads, wings, hair, and all.

In 1965, when Borja was twenty-three and just married, he left his young wife, Cecilia, and went to California, where he got a job picking apples in the Salinas Valley. From there, he joined the U.S. Navy, and Cecilia came to live with him at the base at Long Beach. When Borja retired from the Navy, in 1985, he and Cecilia decided to return to Guam. He had not been back in twenty years. The Borjas settled on Billy's old family farm in Sinajana with the three youngest of their five children. Cecilia got a job as a government clerk, Billy hunted and fished, and they tended a flower garden together. Guam had changed while the Borjas were away. Even before they went to California, most of the thatched huts had been replaced with concrete houses, which would not blow down in the typhoons. Now there were gas stations and shopping centers in Sinajana. The boonies had been invaded by brown tree snakes. All the birds were gone because the snakes had eaten them. The fruit bats were almost all gone, too, overhunted by the Chamorros and the snakes.

One day in October of 2002, Borja, a thin, fit man of fifty-nine, was out cutting brush on the farm. Suddenly, for no apparent reason, he stumbled and fell to the ground. In the next few months, he fell several more times. His wife noticed that he was walking oddly; although his right arm swung freely, his left hung limp. Early in 2003 he developed a shuffling gait and a strange, fixed stare. Borja went to see a doctor at the Guam naval hospital, who referred him to a neurologist named John Steele. "The most striking immediate feature of his appearance is a staring quality of his gaze," Steele wrote in his examination notes. "He blinks very infrequently and when he does, the eyelids close slowly and are slow to open." Borja was able to swivel his eyes to the left and the right but not up or down. Steele had seen that stare before. As a young resident in a hospital in Toronto in the early 1960s, he and two senior colleagues had identified a rare neurological disease called progressive supranuclear palsy, or PSP; the disease is often called Steele-Richardson-Olszewski syndrome, after the doctors who discovered it. In 1983, Steele moved to Guam, where he still lives; over the years, he has encountered dozens of patients with the fixed gaze that is characteristic of PSP victims.

Steele recommended for Borja a drug usually prescribed for Parkinson's disease, but he was not optimistic. At this point, he suspected that Borja had PSP, which is untreatable and implacable. Borja's speech began to slur; he tired easily. He grew weak and could stand only if two of his sons held him up, one on each side. Cecilia quit her job to take care of him, and when he could no longer walk, she and the children drove him wherever he wanted to go. Borja told them that he was frightened. Often, he couldn't sleep, for fear that he might never wake up. One night last August, he grew very agitated. "Why're you whining like a baby?" his wife scolded him gently, although by then he couldn't speak. She reproaches herself now for not taking the moment more seriously—she was in denial, she says. The next morning, while she was feeding him breakfast in bed, he grabbed her arm and pulled it to his chest, something he had never done before. Their youngest son was getting ready for work, and she hurried out of the bedroom for a minute to help him. When she went back in, Borja was dead.

For more than a hundred and fifty years, Chamorros on Guam and on Rota have suffered from a strange neurological disease, which they call lytico-bodig—"lytico" from the Spanish word for paralytic and "bodig" from a Chamorro word for listlessness. The symptoms are polymorphous: some cases present like Alzheimer's disease; others like Parkinson's or amyotrophic lateral sclerosis (ALS); still others like PSP. These diseases all have in common a peculiar pattern of neurofibrillary tangles in the brain, and for this reason some neurologists call them the tangle diseases. At the height of the epidemic on Guam, in the 1950s, the incidence rate was as much as a hundred times as high as the global average for ALS. In the worst-hit village, Umatac, a small, remote place where Magellan is said to have landed in 1521 during his voyage around the world, a Chamorro lay sick in almost every hut.

In the 1950s and 1960s, scientists from around the world traveled to Guam and Rota, hoping that if they could figure out what was happening in the islands, they would have a clue to the cause of some of the world's most poorly understood diseases. In 1956 the National Institutes of Health established a research station on Guam, to look into the cluster of tangle diseases in the Marianas. Clusters of these diseases are usually small; most are statistical flukes. This cluster was by far the biggest on record. It was so pronounced and so geographically isolated that many neurologists assumed the answer would be easy to find. And the reward would be huge: a scientist who managed to unravel the mystery could help save innumerable lives and would very likely win a Nobel Prize and millions of dollars of funding for research. But by the early 1980s, when Steele arrived on Guam, the epidemic was already beginning to recede, the research station had closed, and there was still no explanation for the disease, much less a cure. Since then, there has been a dramatic drop in the number of new diagnoses; the only people at risk seem to be those who, like Borja, lived on Guam before 1950 or so. Whatever triggered the disease seems to have disappeared with the Chamorro way of life.

By the turn of the century, the case had gone cold. Fifty years of research seemed to have accomplished almost nothing. On Guam, Steele and a few others still treated and studied the epidemic's last surviving victims. Elsewhere in the world, scientists remembered Guam as a sort of fable of neurology and its difficulties. Then, three years ago, a newcomer to the problem ventured a guess in the journal *Neurology*, provoking the curiosity and skepticism of nearly every specialist who read it. The author, Paul Cox, was an ethnobotanist; as the director of the National Tropical Botanical Garden on the Hawaiian island of Kauai, and an expert on Pacific fruit bats, he was a complete outsider to neurology. But Cox was convinced that he had succeeded where generations of more qualified scientists had failed.

FOR DECADES, some Chamorros have placed the blame for the disease on the islands' cycads, which look a little like palms or tree ferns. Cycads are the most primitive seed-bearing plants known on earth—they predate the dinosaurs by tens of millions of years—and their seeds, each about the size of a squash ball, are poisonous. To detoxify them, Chamorros chop them up and soak them in pails of water, changing the water daily for a week or two. Then they use the seed pulp to make flour for tortillas and chips. If dogs or chickens drink the water from the first soaking, they die. Even ants and flies stay away from it.

In the 1950s Marjorie Whiting, an ethnobotanist, and Leonard Kurland, an epidemiologist and neurologist, who was the director of the NIH research station, decided that the Chamorros might be right about cycads. Whiting spent a month in Umatac, cooking with the women and recording their recipes. She lived with the Quinata family, who, according to local folklore, were the first people on Guam to come down with the disease, and who had lost someone to it in every generation since the early 1800s. The cycad theory gained popularity, and in 1962 Kurland was able to organize an ambitious series of cycad conferences sponsored by the NIH. Many of the conference sessions focused on cycasin, a toxic compound in cycads. But in laboratory tests cycasin did not seem to kill nerve cells. (It did turn out to be a powerful carcinogen, however.) At a conference in 1967, a pair of British biochemists, Arthur Bell and Peter Nunn, reported that they had isolated another toxic compound in cycad seeds, beta-methylamino-L-alanine, or BMAA, an amino acid. BMAA did kill nerve cells, if it was applied in huge doses in a petri dish. But laboratory rodents that were fed BMAA did not get sick with lytico-bodig, and they did not develop neurofibrillary tangles. In any case, the analysis of cycad flour revealed only trace amounts of BMAA. To poison yourself, it seemed, you would have had to eat several tons of cycad chips.

The cycad excitement died down, Kurland left Guam, and in the early 1980s the research station's next director, D. Carleton Gajdusek, explored a new hypothesis. Gajdusek, who had worked on kuru disease among the cannibals of New Guinea, and had been awarded the 1976 Nobel Prize (with Baruch Blumberg) for studies of infectious diseases, proposed that the disease might be caused by mineral deficiencies that led to a buildup of aluminum in nerve cells. He collected Chamorro brain samples from autopsies and sent them to a young neuropathologist named Daniel Perl, who was then at the University of Vermont. Perl was becoming well known for having found aluminum in the neurofibrillary tangles of Alzheimer's patients—data that helped start a minor panic about cooking with aluminum pots and pans. By the end of the first day of his examination, Perl knew that the neurofibrillary tangles in the Chamorro brain samples were loaded with aluminum, too, and at concentrations about ten times as high as those in Alzheimer's patients' brains. Among some scientists, the cycad hypothesis came to be known as "Kurland's folly."

Soon, however, other research began to suggest that the aluminum in the neurofibrillary tangles of Alzheimer's patients might be an effect rather than a cause of the disease. Nonetheless, Gajdusek asked Perl to be the Guam project's neuropathologist. He sent Perl many of the NIH's Chamorro brain samples, and Perl stored them in what neurologists call a "brain bank." When Perl left Vermont for the Mount Sinai School of Medicine in Manhattan, in 1986, he took the bank with him.

Cox's theory, which was published in *Neurology* in the spring of 2002 (the neurologist Oliver Sacks, who has written a book on the Guam epidemic, was the paper's co-author), drew together many strands of previous research, and connected them neatly with a single factor that seemed to have been overlooked by everyone who had studied the disease: some Pacific fruit bats eat cycad seeds. Cox suggested that Chamorros who ate the bats might have consumed BMAA in concentrated doses. He noted that the peak of the Guam epidemic had followed the influx of American guns and cash to the island in the first half of the twentieth century. Fruit bats roost in trees and are hard to kill without guns, but one shotgun blast can bring down a few at a time. By the mid-seventies, Cox wrote, the bats had become an endangered species on Guam, and the Chamorros were importing them from other Pacific islands—thousands from Western Samoa alone. These imported bats would not have been neurotoxic because there are no native cycads in Samoa. That might explain why the only Chamorros to contract the disease were people old enough to have eaten native bats.

A thickset, charismatic, intense-looking man of fifty-one, Cox has spent much of his life either deep in the wild or deep in the laboratory. In repose, he looks pale and scholarly, but when he grins his face creases with laugh lines like a Western rancher's. He grew up in a Mormon family, in Provo, Utah, and has an easy way of making friends and gripping a shoulder to seal the pact. He has a gift for salesmanship, too, and as soon as his hypothesis appeared in *Neurology* he began barnstorming for it at universities around the world. Cox was well aware that he and his team—two young scientists, who worked with him in a tiny laboratory on the grounds of the botanical garden in Kauai—were venturing into hostile territory. "I'm a botanist—I carry a hand lens," he liked to say, when he lectured to tropical biologists. "My previous expertise is limited to a backpack and a machete." And lytico-bodig was no ordinary problem to neurologists. "As far as I can tell, this is like Fermat's last theorem," Cox often said.

When I first met him, in January 2004, he was in the process of testing the bat hypothesis, and had been compiling data that he thought would rattle the scientific world. By now, he had published a second paper about his theory in *Conservation Biology;* the journal's cover showed a color photograph of a bat with its mouth stretched wide around a cycad seed. But he told me that he was trying to keep his latest work quiet until he had published enough papers so that he wouldn't be ridiculed. "I mean, come on, we're gonna get blown out of the water. Our little Mickey Mouse program! When we publish these papers, the response will be shocking and immediate. There will be five hundred reasons why we're wrong. Some of which might be true. Maybe we are wrong. It's a very scary business. These neurologists have such big egos. They're the Porsche crowd. I like flowers. I live in remote villages. Botanists, we're pretty much gentle folk."

Cox first glimpsed the South Pacific's fruit bats when he was nineteen, and was in Samoa on a mission for the Mormon Church. The bats circled above the rain forest like eagles; at sundown, they squabbled in the treetops, drowning out the sounds of the village. They had no fear of human beings. The village chiefs taught Cox the language of orators, and he learned to declaim at village ceremonies, wearing a formal lavalava and a red lei and bearing a ceremonial staff made from a kava root. A year into his mission, he fell sick with a high fever. While he shivered and shook, a village healer rubbed roots on his head and chest. "That and prayer healed me," he says, with the kind of cheerful laugh that scientists who are religious learn to use when talking about faith with rationalists.

After his mission, Cox studied botany at Brigham Young University, and then ecology and evolution at Harvard; his Ph.D. thesis was on the ecology of the Samoan rain forest, including the bats. He won Harvard's Bowdoin Prize for essay-writing twice: first for a paper about plant evolution, and then for one, written under the pen name Bat Masterson, about plant evolution and Samoan bats.

In 1984, when Cox was an assistant professor at Brigham Young, his mother died from breast cancer. He considered becoming an oncologist, but decided that, rather than help a finite number of people, he would use his knowledge of rain-forest ecology to try to find a cure for breast cancer. The following year, with his wife, Barbara, and their four children, he returned to Samoa to collect plant specimens. He also continued to observe fruit bats. In a 1997 memoir of his years in Samoa, he describes a bat that hovered just over his head and stared down at him, high on a ridge above the Pacific. He writes, "To have such a gentle but wild creature scrutinizing you under no constraint other than its own curiosity is to feel the weight of God bearing upon your soul."

It was in Samoa in the mid-eighties, Cox says, that the bat hypothesis came to him. He did nothing with the idea at the time. "I was just too timorous," he says. He returned to Brigham Young and pursued a promising therapy that a healer in the small village of Falealupo had shown him: an extract from the bark of the mamala tree, which Samoans used to treat hepatitis. Cox brought a sample of the bark to the NIH, where scientists found that it contained a chemical that caused a cell infected with hepatitis to eject the virus. A nonprofit research group, the AIDS ReSearch Alliance, is now developing the chemical, prostratin, for use with AIDS patients. Last fall, Cox helped broker a deal between the University of California at Berkeley and the Samoan government. Scientists at Berkeley will try to isolate key genes from the mamala tree; any profits will be shared with the people of Samoa.

Cox has enjoyed a good deal of professional success as an evolutionary theorist and botanist. At Brigham Young, he served as the dean of general education and honors from 1993 to 1997. He has also been an effective conservationist, helping to create a national park in American Samoa; protecting its fruit bats from being exported to Guam as a delicacy; and establishing a foundation that helps indigenous healers and shamans around the world, in return for their setting aside nature preserves on and around their territories. For one conservation project, Cox and a Samoan chief received, in 1997, the Goldman Environmental Prize, which is sometimes called the conservationist's Nobel. In Samoa, the village chiefs gave him one of their most distinguished titles, Nafanua—protector of the rain forest and deity of war.

Cox has also made enemies along the way. Some biologists say that he is a publicity hound; they call him "the Carl Sagan of Pacific botany." He is also known for being highly competitive: he once threatened to sue a scientist who criticized a favorite student's work. At Brigham Young, students repeatedly voted him Teacher of the Year, but he angered some of his colleagues by pressing hard for his causes—for example, insisting that the department consider hiring one of his protégés—and by taking many long leaves in Samoa. "Paul has a silver tongue," one former Brigham Young administrator told me. "He's a very self-serving individual. He caused me more consternation than all the rest of the faculty put together by a factor of about a hundred. He'd go over people's heads—right to the board of directors. A lot of people don't have a lot of love in their hearts for the guy."

Gary Booth, an environmental toxicologist and Cox's closest

friend at Brigham Young, defends him, saying that he is honest and diligent. "I never saw any indication that he was trying to snowball or browbeat anyone," Booth said. He thinks that Cox's colleagues were just jealous. "Seems to be, either you like Paul or you don't. There's not too much middle ground." Another supporter, a neurologist, acknowledges that Cox is driven by something uncommon among scientists. "I didn't know Paul had been a Mormon missionary," he said. "But I have noticed in him"—he paused and gave me a small smile—"an impulse toward salvation."

In 1997 Cox left Brigham Young to become King Carl Gustaf XVI Professor of Environmental Science at the Swedish Agricultural University in Uppsala. The following year, he took up the position of director of the botanical garden in Kauai, and began to work on the bat theory that had first occurred to him a decade before. After publishing his hypothesis in *Neurology,* Cox set out to find evidence to support it. His goal was to trace cycad neurotoxins through the Guam ecosystem: from the plants to the bats and into the brains of Chamorros who had died of the disease. With no advanced training as a biochemist, he had to teach himself. "I'm paddling in weird amino acids, and it feels like I'm drowning," he says of that time. "My wife's afraid the chemistry books are gonna collapse on my head if there's an earthquake." Eventually, he realized that he needed to collaborate with someone who had more experience, and he brought in Susan Murch, a plant chemist at the University of Guelph in Canada. Murch was as much an outsider to neurology as he was, but between college and graduate school she had spent five years as a chemist at the Hospital for Sick Children in Toronto, doing amino-acid analyses in an experimental program for premature babies. Cox also invited Sandra Banack, an ecologist with a special interest in Pacific bats, who had been his first graduate student at Brigham Young, to join the small research team.

In the spring of 2002 Cox and Banack got permission from the Museum of Vertebrate Zoology at Berkeley to examine three bat specimens that had been collected in Guam in the 1950s. They cut off bits of dried skin, with hair attached, and took them in sample bottles to Kauai. Banack worked around the clock, feeding the samples through a high-performance liquid chromatograph, a chemical analyzer that sorted the molecules in each sample. One morning at two o'clock, the results scrolled across the machine's computer screen in a series of jagged peaks. If her measurements were correct, one of the peaks represented the chemical marker for BMAA. It implied that the bats' flesh contained thousands of times more nerve poison than the cycad seeds: a single bat held as much BMAA as a ton of cycad flour. "It's like your heart stops," Banack told me. "I just looked at it and thought, Oh, my gosh, we're right. I couldn't believe it. Then I picked up the phone and called Paul. He answered the phone so fast I knew he was up. And we just both sat there on the phone in silence."

Next, Cox collected cycad samples from Guam, from his botanical garden in Kauai, and from two other gardens in Florida. He and Banack tested cycad seeds and their outer skins—the only part of the seeds that a bat eats. According to their measurements, the outer skin seemed to have a far higher concentration of BMAA than the rest of the plant. They also checked for BMAA in the cycads' roots. Many species of cycad live symbiotically with cyanobacteria, blue-green algae that invade some roots and redirect their growth to the surface. Botanists call these wrong-way roots coralloid: they look like tiny gnarled pieces of coral in the soil around the cycad's trunk. When you break a coralloid root, you can see a shiny green layer of cyanobacteria, almost the color of kiwi fruit. Cox and his team ran ten root samples through the chromatograph, and concluded that the blue-green algae manufacture the BMAA and spread it through the plant and into the seeds.

The next step was to check for BMAA in Chamorro brains, which Cox obtained, with John Steele's help, through Patrick McGeer, a neuroscientist in Vancouver. McGeer sent Cox brain samples from six Chamorros who had died of lytico-bodig, and two who had died of other causes. Bats and cycad chips are so popular among traditional Chamorros that it is hard to find anyone who hasn't eaten them, so

McGeer also sent Cox and his team brain samples of Canadians from the Vancouver brain bank as controls: thirteen who had died of causes unrelated to neurodegeneration, and two who had died of Alzheimer's. These tissue samples were pickled in paraformaldehyde. The study would be "blind"—the identity of the samples would be concealed until the end of the experiment.

This time it was Susan Murch who was alone in the lab when the rows of peaks appeared on the chromatograph's screen. When the blind was broken, she saw that, according to her measurements, the six Chamorros who had died of the disease had BMAA in their brains. Murch also found BMAA in the North American samples—but only in the two Canadians who had died of Alzheimer's. "I can't tell you how shocking and unexpected that was," Cox says. "I started weeping."

In August 2003 Cox and Banack reported in *Neurology* their discovery of high concentrations of BMAA in the bat tissues and in the cycad seeds. He told the botanical garden's trustees about his findings, and one of them bought the laboratory a second, more expensive chemical analyzer, a mass spectrometer, which Cox and his team used to recheck their measurements. Soon they made another surprising find. They detected BMAA not only as a free compound but also in bound form. According to their data, the BMAA seemed to be doing more than just floating around in the tissues of the cycads, the bats, and the human-brain samples. It appeared to have become part of the tissues, incorporated into the chemical structure of the living proteins. To Cox, bound BMAA represented a crucial piece of evidence for his hypothesis. He speculated that it might help cause the tangles that are characteristic of lytico-bodig, and that it might even explain the tangles of Alzheimer's, Parkinson's, and other neurodegenerative diseases. It could also explain lytico-bodig's long latency period. "That's key, finding it in protein," Cox told me. "If you claim an environmental neurotoxin is causing lytico-bodig in Chamorros, you've got to explain why just Chamorros and no others on the island get the disease. Bats could explain that. Second, why the latency period?

Chamorros can leave and come down with it fifteen years later in California. Why? Well, the protein story builds us a lot. Here's a poison that won't cause acute toxicity. If you've got it in your body, you're getting slowly dosed." This point held a morbid fascination for Cox, and he kept coming back to it. "Your own body becomes a neurotoxic reservoir," he told me.

Cox and his team believed that they had now followed the BMAA from the cyanobacteria in the cycad roots, to the cycad seeds, to the bats, and to the brains of Chamorros who ate the bats. These findings were published in the *Proceedings of the National Academy of Sciences.* It looked to Cox like a classic biomagnification story: on Guam, Cox argued, the bats seem to have concentrated the cycads' neurotoxins the way algae-eating fish concentrate PCBs, so that by the time the bats had been consumed by Chamorros they were carrying high doses of the poison. Elsewhere on the planet, where people do not eat bats, they might get concentrated doses of BMAA from cyanobacteria in other food in their diets, or from contaminated drinking water.

Cyanobacteria are the oldest living things on earth. They may go back as many as three and a half billion years. Through photosynthesis, they gave us the oxygen we breathe. Cox began to wonder if they might also be the primal enemy of creatures with nervous systems. By now, Cox was keeping a plastic model of BMAA in his knapsack, and another by his bed. "The molecule from hell," he called it. "It's too bad, because it's a pretty thing," he said. "It's just causing so much suffering, so much agony among the Chamorros—and maybe us also. I'm not a chemist or anything, but this is a bad molecule." Cox knew that he was getting ahead of himself, but he was willing to let his imagination run wherever the story seemed to point. And he thought he could explain why such a common neurotoxin so rarely causes clusters of disease. Cancer researchers talk about gene-environment interactions—some smokers develop lung cancer, while others can smoke without becoming sick—and Cox speculated that there might be a similar genetic vulnerability at work with blue-green algae. The Chamorros of Guam, especially those in remote villages like Umatac, where inbreeding has been, at times, a reproductive necessity, might carry one or more genes that made them susceptible.

Cox's hypothesis cut across many fields, and in every one of them there were specialists who thought that he had built a big story out of very little evidence. At the University of Guam, for instance, Thomas Marler, a cycad expert who grew up on the island, read Cox's papers and was surprised to see that Cox and his team had put together their picture of cycad BMAA after testing only a few plants. Marler had been working for several years on a study of the plants' biochemistry, involving hundreds of cycads and meticulous sample measurements. He wondered how these scientists could measure the outer layers of three seeds or a single stem and say anything about the typical biochemistry of cycads. Besides, Cox's "typical" cycad was a composite drawn from several plants—from Florida, Guam, and Kauai. "That's just not science," Marler told me. In a paper that is soon to be published in *HortScience*, a journal of the American Society for Horticultural Science, Marler reports that, according to his study, cycad biochemistry fluctuates dramatically from plant to plant and from season to season. He calculates that a sample size of three would give wrong results about ninety per cent of the time. A sample size of one would be wrong 97 percent of the time.

The bat BMAA data that Cox published came from only three museum specimens, and in the same issue with Cox and Banack's paper, *Neurology* ran an editorial by an ALS expert who suggested that the bats might have been atypical, even sickly: "The high concentration of BMAA in the museum specimens may have been the reason why these specific bats became museum exhibits." Cox's announcement of BMAA in human brains was also based on a small number of samples. Douglas Galasko, an Alzheimer's expert at the University of California at San Diego, who runs a Guam research program with funding from the NIH, told me that a typical Alzheimer's study would include larger sample sizes. "In the Alzheimer's field, there've been a large number of claims of environmental causes, ranging from spirochetes and viruses to chlamydiae, a form of organism between viruses and bacteria," he said. "Many of these reports were based on microscopic examination of autopsy material which did not hold up to repeated scrutiny." Galasko also pointed out that the specimens had been pickled in paraformaldehyde, which transforms the chemistry of the tissue it preserves. Measuring amino acids in paraformaldehyde-fixed tissue is not standard procedure and could easily lead even the most experienced chemist astray. "It's just not done," Galasko said.

Galasko conceded that there were once large numbers of Guamanian fruit bats, that some Chamorros ate the bats, and that both the bat population and the epidemic are now in decline; but he said that Cox had not proved a connection. "As a silly analogy, take Spam, which is widely eaten on Guam," Galasko said. Spam was available only after the Second World War, and after Spam-eating rose, in the 1950s, the epidemic began to fade. "Well, perhaps Spam is protective against the disease!" Galasko said. "That's the problem with correlations, you see—they do not prove causation." He reminded me that scientists have fed BMAA to animals and have yet to see anything that looks like lytico-bodig. (Investigators are studying some of the other toxins in cycads, including cycasin and steryl glucosides, which some see as much more likely culprits than BMAA.) He also doubts that BMAA could bind to protein as neatly as Cox proposes. Cox's theory, he says, is elegant but unrealistic—"a nice human-interest story that stirs great sympathy."

Cox's historical data have been criticized, too. The centerpiece of the *Conservation Biology* paper is a graph that shows the population of Guam's fruit bats standing at sixty thousand in 1924, then crashing to a few hundred around 1960 and to near zero by 1984. Cox arrived at the figure of sixty thousand by using several sources, including a paper by the leading expert on Guam's bats, Gary Wiles. When I called Wiles, he told me that he had warned Cox's group that the figure was only a very rough estimate. And although Wiles has found cycad pulp in the bats' scat, other naturalists doubt that they eat cycads often. According to Anne Brooke, a government wildlife ecologist on Guam, the cover photograph in *Conservation Biology* was posed. It was taken by Merlin Tuttle, of Bat Conservation International, in Wiles's apartment. "He had a captive bat and stuck fruit in its mouth," she said. The bat was hungry; if Tuttle had given it a chocolate bar, the bat would have eaten that instead. "Every time Paul Cox comes up with another award or a big glossy story about him, we all just cringe," Brooke said. She and her colleagues on Guam can rattle off many other objections to Cox's hypothesis. For instance, there is a small village next to Umatac called Merizo. Chamorros in Merizo loved to eat bats, too, and yet the disease rate in Merizo was five or six times lower than in Umatac.

John Hardy, an NIH scientist who studies the genetics of neurological diseases, including lytico-bodig, told me that he believed the bat story appealed to Western stereotypes about the Pacific Islands. "I think the evidence is really zero," Hardy said. "Also, the allure of the story is subtly racist—it attracts the popular imagination because it sounds so weird. It's really kind of condescending. We all eat funny things. I'm from North Lancashire, and I eat black pudding. The French eat sparrows. The Chinese used to think we were crazy for eating milk and cheese. We take for granted our own habits and think others are bizarre. We have to be careful we're not just falling for this exotica view."

Cox's most caustic critic is Daniel Perl, the keeper of the Chamorro brain bank. Perl is a slight man in his early sixties, with a salt-and-pepper mustache. When I met him for lunch last fall at a restaurant near Mount Sinai, he wore a dark-blue blazer, and a doctor's ID was clipped to the pocket of his button-down shirt. "I don't want to sound like a disgruntled competitor," he said, after we sat down. "But I think some people who have gotten into this business are swinging for the fences, making statements they have no business making."

Perl told me that he thought Cox's science was generally irresponsible, and would make it only harder to fund fresh research projects in the Marianas. "Anyone who works on Guam—you never lose interest," he said. "It's always in the back of your mind. Kurland said it best: It's like the most fantastic murder mystery you ever read, and you just want to know who did it. And sadly, toward the end, all Kurland cared about was that he wanted to find out while he was still alive. And now I'm beginning to say that. I say from the bottom of my heart"—he stretched out his arms theatrically—"I don't care who solves it. I just want someone to explain it to me. But, along the way, I don't want the path obscured by nonsense! This is a hard enough problem without that." Perl, collaborating with Galasko's NIH-funded group, was trying to replicate Cox's BMAA measurements. He doubted that they would find any.

"I'm not sure what they're detecting, okay?" Perl said. "My guess is they're detecting something else. God knows what it is."

For Cox, with so many questions hanging over his research, the sensible thing would have been to stay in the lab and try to replicate his contested results. But he did the opposite. He was eager to show that his findings on the island applied all over the world. "I don't mind being wrong and going down in flames," he told me. "I've had a good enough scientific career that if I flame out that's fine. For me, hell would be that I'm right, but, because of my own inability to figure it all out, the case is not proved for another fifteen or twenty years. My grail now is to raise this story to the level of scientific respectability, so guys with major funds come in—so a few big guys come in with the goal of swatting us."

Last spring Cox set off with Murch on a BMAA-collecting trip around the world. In the years since Western doctors became aware of the epidemic on Guam, they had learned of similar outbreaks in Irian Jaya in Indonesia; on Japan's Kii Peninsula, southeast of Osaka; and on Guadeloupe in the Caribbean. Cox thought his findings might turn out to explain those outbreaks, and others, too. He and Murch trekked on skis with reindeer herders in Lapland: he wanted to see if BMAA could enter the ecosystem through the lichen that the reindeer eat, some of which are rich in cyanobacteria. He went to China to collect samples of "Lucky Soup," which is made from blue-green algae, and surveyed them in a lab in Zurich, inspecting chain after chain of algae like strands of poison pearls. He bought samples of virtually every food for sale in some of the Kii Peninsula's most remote villages, where the incidence of ALS is now as high as the incidence of lytico-bodig ever was on Guam.

When I visited Cox on Kauai late last spring, he had just returned from Japan. He met me at the airport and put me up at a guesthouse on a beach at the edge of the botanical garden. The next morning, on my way to the lab, I hiked through one of the garden's tropical valleys, stopping to admire waterfalls, lava cliffs, green hills of palms and cycads, and a grove of giant eucalyptus trees. (Steven Spielberg filmed scenes from *Jurassic Park* in that valley.) When Cox found me in the lab, I was staring out the window.

"Sorry about the hellhole I've dropped you into," he said cheerfully.

The lab was well equipped: the chromatograph and the mass spectrometer were both the latest, most sophisticated models. Two big freezers held samples of Chamorro brain tissue that Steele and McGeer, the Vancouver scientist, had shipped to the lab the previous summer; Cox was just getting around to testing them.

At the laboratory bench, Susan Murch wore a white mask and lab glasses and a lab gown. The lab technician, Holly Johnson, wore lab dress, too, except for a pair of flip-flops. The scientists were working their way through their Chamorro and Canadian brain samples, along with vials containing Chinese soup, Japanese fish, and reindeer fur, which Cox and Murch had brought back from their trip. The smell of the human brains in the vials was faintly putrid.

Cox and his crew were working long hours, but it was slow going, and it was clear both to them and to me that they were seriously overstretched. Each of their BMAA data points required an elaborate series of procedures involving vortexing, heating, immersing in acid baths and in liquid nitrogen, and freeze-drying. It took them several days to prepare each brain sample for measurement. Then the vials went one by one into a carousel of the chromatograph, rather like slides in the tray of a projector. At last, after a dramatic pause, six different lines—umber, purple, lime-green, sky-blue, lavender, and dark brown, each representing a different sample—began inching across the instrument's computer screen from left to right. Most of the samples were clean, but a few of them generated a pronounced spike that to Murch indicated the presence of BMAA. She showed me how she double-checked each of her results with the mass spectrometer, which analyzes compounds by blasting the molecules apart and measuring the charges of the fragments.

Before my visit, Cox had asked me to send him a lock of my mother's hair so that he could demonstrate his methods. My mother had been given a diagnosis of PSP, the disease that Steele helped discover in 2000. By last spring, when I sent Cox her hair, she was almost completely paralyzed, just barely able to swallow puréed foods from a spoon; she could not speak, and her face had become a frozen mask. This January, after years of suffering, she died.

In the lab, I handed Cox two baggies, with locks of hair from two other close relatives. For some reason, I had not got around to cutting a lock of my own.

"Let me do the butchering," Cox said, and made a few snips.

Murch took each bag, numbered it neatly with a pink Sharpie, and entered the number into her lab notebook. Then I entered her numbers into my own notebook, with my own identifications, which I did not let her see. Cox added a few other hair samples, from the freezer. Murch knew that one of the samples in the run was my hair. She didn't know that three other samples came from my family, nor did she know the identity of the two other samples that Cox gave her. But one of those came from a close mutual friend of theirs with brilliant auburn hair, and the other sample was just as distinctive. (Cox has asked me not to describe it.) So for this round, at least, the blindfold was a bit loose.

A few days later, the hair was ready for testing. Murch put each sample into the chromatograph's carousel. The results for five of the six samples were clean: she saw no BMAA. But when Murch ran sample No. 1528, the line formed a sharp peak. She rechecked that sample in the mass spectrometer, with the same results. According to her measurements, that peak was the chemical signature of BMAA.

I broke the blind to reveal that No. 1528 was my mother's hair. Cox said, levelly, "She's got it." He looked grimly satisfied. "As scientists, we can't go beyond our data," he cautioned me. "All we can say is, a molecule that occurs in victims of ALS in Guam has also accumulated in your mother."

The next morning, Cox found me walking in the garden. He rolled down the window of his Jeep. "Get in. Something exciting happened," he said. He told me that he had been looking at the computer that morning and realized that, with the BMAA measurements from my family's hair, his data had reached the level of statistical significance. "That did wonders for my spirits. I don't mean to glory in your personal anguish, but clearly there's a signal here."

BY LAST FALL, Cox had decided to devote himself to BMAA research. As he prepared to step down as director of the botanical garden, some of the garden's trustees and benefactors established a nonprofit foundation, the Institute for Ethnomedicine, to fund his project, and Cox planned to build a new lab back in Provo or in Wyoming. "By a year from now, I want to be pretty deep into neurochemistry," he told me. "See if I can find some way to protect the nerves. It's a risky move. A more rational academic move would be to sit and tie up all the loose ends of what we're doing now. But, again, I want to shoot as fast as possible toward therapy."

His BMAA work had found supporters as well as critics. Raymond Roos, a professor of neurology at the University of Chicago School of Medicine and one of the most measured Guam watchers I spoke with, said, "I admire Paul because he has a new hypothesis and it's a testable one and he's trying to test it. He's an ethnobotanist, so he's got to interact with the right people and do the right experiments. But he's got a lot of experience moving here and there and talking with all kinds of exotic people, so hopefully he'll be successful in doing that."

Cox began collaborating on a separate set of tests by a well-respected neurology group at the University of Miami, supervised by Deborah Mash, who runs the brain bank there. Cox's Institute for Ethnomedicine bought the university a high-performance liquid chromatograph, and Mash's group planned to collect samples from ALS patients, compare them with samples from their spouses, and measure the levels of BMAA in ALS patients against control groups. "I'd say the jury's out," Mash said of the research. "But I'd bet on Cox. He's one of the most fiercely intelligent people I've met in a long time." Walter Bradley, an ALS expert and the chairman of the university's neurology department, was enthusiastic. "If Cox pans out, he probably will get the Nobel Prize for this—or should," he said. "It really is a brilliant piece of detective work."

This week, after slowly building his case in papers published during the past three years, Cox, with Murch, Banack, and several other scientists, plans to announce in the *Proceedings of the National Academy of Sciences* that he has found BMAA in species of cyanobacteria that grow all over the world—from the United States to Scotland, Israel, Australia, and Zanzibar, and from the Baltic to the Sargasso Sea. Based on measurements of thirty cyanobacterial samples, Cox and his coauthors suggest that BMAA may link the local problem of lytico-bodig to the global problem of neurodegenerative diseases. If Cox turns out to be right, the blooms of blue-green algae that sometimes cover thousands of square miles of ocean or taint the reservoirs that hold our drinking water may contain a potent neurotoxin. We may risk consuming it wherever we live, whenever we eat a piece of fish or take a sip of water.

In the last days of November, Paul Cox invited a group of scientists to a conference on BMAA in Park City, Utah. Cox had been talking and visiting with all of them, but most had never met each other. One of the world's leading experts in cyanobacterial toxins,

Geoffrey Codd, was there from the University of Dundee, where he had been hosting a conference on the global health hazards posed by poisonous algal blooms, mats, and scum. Arthur Bell, who had led the team that isolated BMAA in 1967, and had gone on to become the director of Kew Gardens, came in from London. So did Bell's colleague in the BMAA discovery, Peter Nunn, and Peter Spencer, who led a famous but short-lived revival of the cycad hypothesis in the 1980s. Nunn and Spencer had recently attended a conference in Syria on lathyrism, a kind of paralysis caused by eating too many chickling peas. (Spencer has been studying cycasin, the other notorious toxin in cycads.) There were three biologists from Stockholm, and two from the City of Hope, a biomedical research center in Southern California. The two from California reported on their tests of BMAA's effects on fruit flies, experiments they had begun in response to the Guam story. "We're newcomers here," they said.

Cox arranged for the group to be ferried in vans from the conference to a private house for dinner. The house had cathedral ceilings and cathedral-size beams, and it was decorated for the season with holly wreaths and Christmas lights. Outside, there was a foot and a half of new snow. The group settled down around the fire. Oliver Sacks spoke poetically about the extraordinarily long Guam quest. People took turns recalling Marjorie Whiting, Leonard Kurland, and others from the first wave. More than one of the specialists in the room spoke of feeling that an immense jigsaw puzzle was beginning to come together. "The story may be much more complex than we can see from here," Cox said. "BMAA might just be the flag sticking up from the parade of a whole cocktail of nasties coming at us." But he felt more hopeful than ever that BMAA would prove to be a crucial clue.

Cox had invited only supporters to the conference. Meanwhile, Galasko and Perl's NIH-funded group, which had been trying for months to replicate Cox's work using a different set of brain samples, was about to unblind its study. They were unable to find any BMAA in the brains of patients who died from tangle diseases—neither in those from North America nor in Chamorros from Guam—and they have prepared a paper for publication. On the other hand, Mash and her team in Miami tell me that although it is too soon for them to publish, their early BMAA results do match Cox's.

At the close of Cox's conference, the mood was guardedly optimistic. When the circle broke up, I sought out John Steele. He was wearing a navy-blue blazer and tie; his face was ruddy to the roots of his white hair. Steele has seen the cycad hypothesis wax and wane a few times before, and he has learned to beware of the lure of a good story. Each time, however, he cannot help but be drawn in. "I come from a family that suffers from these diseases," Steele had once confided to me. His grandfather and his mother died from Alzheimer's; his sister suffers from Parkinson's; and a few days before the conference his brother had died from Alzheimer's, too. He had flown in from the funeral in Toronto. "Understand one of these diseases, you're likely to understand them all," he had told me. His children have often begged him to move on, but he finds himself unable to, particularly now: "I've seen success just beyond the tips of my fingers, and now I think I can see it again."

Before we left Utah, Steele told me about a Chamorro folktale that traces lytico-bodig to a kind of original sin. According to the tale, Father Cristóbal Ibáñez de San Onofre, an Augustinian priest who came to the island in 1771, placed a curse on the Quinata family. Next to his church in Umatac was a mango tree, from which the villagers were forbidden to harvest fruit. The Quinata family ignored the edict, and Father Cristóbal punished them by afflicting their descendants with the disease. People in the village told Marjorie Whiting that the Quinatas were still stealing those mangoes and selling them in the market, and that was why Guamanians still had the disease—not only the Quinata family but everyone who bought their mangoes.

Steele said that he had once visited the overgrown ruins of the church. Hacking away at the brush with a machete, he found a sprawling old mango, dying, but with new trees sprouting up around it.

David Quammen

Clone Your Troubles Away

FROM *HARPER'S*

Cloning animals, now a reality, is increasingly being advocated as a way of rescuing endangered species from extinction. Investigating this seemingly elegant remedy, David Quammen begins to wonder if cloning really offers a solution to the problem, or whether it's just a form of high-tech wishful thinking.

One morning early last winter a small item appeared in my local newspaper announcing the birth of an extraordinary animal. A team of researchers at Texas A&M University had succeeded in cloning a whitetail deer. Never before done. The fawn, known as Dewey, was developing normally and seemed to be healthy. He had no mother, just a surrogate who had carried his fetus to term. He had no father, just a "donor" of all his chromosomes. He was the genetic duplicate of a certain trophy buck out of south Texas whose skin cells had been cultured in a laboratory. One of those cells furnished a nucleus that, transplanted and rejiggered, became the DNA core of an egg cell, which became an embryo, which in time became Dewey. So he was wildlife, in a sense, and in another sense elaborately synthetic. This is the sort of news, quirky but epochal, that can cause a person with a mouthful of toast to pause and marvel. What a dumb idea, I marveled.

North America contains about twenty million deer. The estimate is a rough one (give or take, say, five million), since no one could ever count them. Some biologists suspect that the number is higher now than it was five hundred years ago, reflecting the impacts of European settlement on the American landscape. Predators eradicated, old forests cut or thinned, more second growth, more edges and meadows—these changes are happy ones for deer. By any measure we've got plenty, and they're breeding like gerbils, poaching lettuce from suburban gardens, overflowing onto highways to become roadkill. Of the two species, mule deer and whitetail, the whitetail *(Odocoileus virginianus)* is more widely distributed and abundant—more abundant, in fact, than any other large wild mammal on the continent. Given such circumstances, it struck me as odd that someone would use postmodern laboratory wizardry to increase the total. Odder still to increase it, at some considerable cost, by just one. Cloning is expensive. Deer, I imagined, in my ignorance, are cheap.

The news item, drawn from wire services, was only a column filler that didn't offer much detail. It barely alluded to the central question: *Why* clone a deer? It mentioned that Dewey had been born back in May, seven months earlier, his existence kept quiet pending DNA tests to confirm his identity as an exact genetic copy. That settled, he could now be presented to the world. Dr. Mark Westhusin, of the College of Veterinary Medicine at Texas A&M, spoke for the team that created the fawn, explaining fondly that Dewey had been "bottle-fed and spoiled rotten his whole life." The item noted that A&M, evidently a leading institution in the field, had now cloned five species, including cattle, goats, pigs, and a cat.

One other claim in this little report (which had the flavor of a reprocessed press release) went unexamined and unexplained: "Researchers say the breakthrough could help conserve endangered deer species." Seeing that, I began planning a trip to Texas.

THE NOTION THAT CLONING might help conserve endangered species has been bandied around for years. Very little such bandying, though, is done by professional conservationists or conservation biologists. One lion biologist gave me a pointed response to the idea: "Bunkum." He and many others who study imperiled species and beleaguered ecosystems view cloning as irrelevant to their main concerns. Worse, it might be a costly distraction—diverting money, diverting energy, allowing the public to feel some bogus reassurance that all mistakes and choices are reversible and that any lost species can be re-created using biological engineering. The reality is that when a species becomes endangered its troubles are generally twofold: not enough habitat and, as the population drops, not enough diversity left in its shrunken gene pool. What can cloning contribute toward easing those troubles? As for habitat, nothing. As for genetic diversity, little or nothing—except under very particular circumstances. Cloning is copying, and you don't increase diversity by making copies.

Or do you? This assumption, like the one about cheap deer, turns out to merit closer scrutiny.

The people most bullish on cloning are the cloners themselves, a correlation that's neither surprising nor insidious. They don't call themselves "cloners," by the way. Their résumés speak of expertise in reproductive physiology and "assisted reproductive technologies," a realm that stretches from human fertility medicine to livestock improvement, and includes such tasks as in vitro fertilization (IVF, as it's known in the trade), artificial insemination (AI, not to be confused with artificial intelligence), sperm freezing, embryo freezing, embryo transfer, and nuclear transfer (which refers to the information-bearing nucleus of a cell, where the chromosomes reside, not the energy-bearing nucleus of an atom). There's also a process called ICSI (pronounced "icksy"), meaning intra-cytoplasmic sperm injection, helpful to elderly gentlemen whose sperm cells can no longer dart an egg with the old vigor. The collective acronym for all such assisted reproductive technologies is ART. To its practitioners, cloning is just another tool in the ART toolbox.

These ARTists are smart, committed people. Like others who feel a vocational zeal, they do what they believe in and believe in what they do. Blessed is the person so situated. But in their enthusiasm for cloning research, in their need to justify their time and expenditures to boards of directors, university deans, or the public, they send their imaginations to the distant horizon for possible uses and rationales. See what cloning could do for you, for society, for the planet? Some of the applications they propose are ingenious and compelling. Some are tenuous and wacky. Three of the more richly peculiar ones, each fraught with complexities and provocations, are cloning endangered species, cloning extinct species, and cloning pets. College Station, Texas, home of Dewey the duplicate deer, is where I picked up the sinuous trail that interconnects them.

"So this guy brought these testicles to me," says Mark Westhusin, as we sit in his office at Texas A&M's Reproductive Sciences Laboratory on the edge of campus. The testicles in question, he explains, came from a big whitetail buck killed on a ranch in south Texas. The fellow had got hold of them from a friend and, intending to set himself up as a "scientific breeder," hoped that Westhusin could extract some live semen for artificial insemination of his does.

Westhusin, an associate professor in his mid-forties, is an amiable man with a full face and a fashionably spiky haircut. He has already explained to me about "scientific breeders," the term applied to anyone licensed by Texas for the husbandry of trophy-quality deer. Deer breeding is a serious business in Texas, where the whitetail industry accounts for $2.2 billion annually, and where open hunting on public land is almost nonexistent, because public land itself is almost nonexistent. Most deer hunts here occur on private ranches behind high fences, allowing landowners to maintain—and to improve, if they wish—their deer populations as proprietary assets. Texas contains about 3.5 million whitetails, some far more valuable than others. An affluent hunter, or maybe just a passionate one, might pay $20,000 for the privilege of shooting a fine buck. A superlative buck, a giant-antlered prince of the species, can be worth $100,000 as a full-time professional sire. And the market doesn't stop at the Texas border. Westhusin has heard of a man who had a buck—it was up in Pennsylvania or someplace—for which he'd been offered a quarter million dollars. He didn't take it because he was selling $300,000 worth of that buck's semen every year. Such an animal would be considered, in Westhusin's lingo, "clone worthy."

Now imagine, Westhusin tells me, that they're collecting semen from that deer one day, and the deer gets stressed, and it dies. Damn. So what do you do? Well, one answer is that you take cells from the dead buck and then clone yourself another animal with the same exact genotype. While you're at it, you might clone four or five.

"You're certainly not going to go out and clone any old deer just for the sake of cloning it," he says. Then again, when you're practicing—when you're developing your methods on a trial basis—you won't wait for the Secretariat of whitetails. The buck from south Texas, the one whose testicles landed in Westhusin's lab, wasn't superlative, but it was good, and the experiment evolved haphazardly.

Working with his students to extract the semen, Westhusin suggested also taking a skin sample from the buck's scrotum, on the chance they might find a use for it. "We'll grow some cells," he said, "and maybe, later on, if we have the time and the money, we'll do a little deer-cloning project." Eventually the effort produced a few dozen tiny embryos, which were transferred into surrogate does, resulting in three pregnancies, one of which yielded a live birth. That was Dewey, born May 23, 2003, to a surrogate mother known as Sweet Pea. The donor buck remained nameless.

The fellow who brought in the testicles remains nameless, too, at least as the story is told by Mark Westhusin. "People don't want it to get out that they've got these huge, huge deer on their ranch. Because then the poaching gets so bad," Westhusin says. Down in south Texas, people circle their land with high fences not just to keep the deer in but to keep the poachers out.

Dr. Duane C. Kraemer, a senior scientist and professor at the Texas A&M veterinary college, is also sometimes called Dewey, though not by visiting journalists or staffers reluctant to presume. He's a gentle, grandfatherly man with pale eyes and thinning hair, whose casually dignified style runs to a brown suit and a white pickup truck. His ART specialty is embryo transfer, and that's the step he oversaw on the deer-cloning project.

Kraemer was the mentor of Mark Westhusin, who did his doctorate at A&M and then worked for several years in the private sector before returning as faculty. The relationship between academic reproductive physiologists and the livestock business tends to be close, even overlapping, because this is a practical science. There's money in assisting the reproduction of elite bulls, cows, and horses, and that money helps fund research. Kraemer himself, raised on a dairy farm in Wisconsin, has been at A&M for much of the last fifty years, during which time he took four degrees, including a Ph.D. in reproductive physiology and a DVM, and performed the first commercial embryo transfer in cattle.

Working on yellow baboons, a more speculative project with implications for human medicine, he did the first successful embryo transfer in a primate. He also did the first embryo transfer in a dog and the first in a cat: Embryo transfer isn't synonymous with cloning—the embryo being transferred needn't be a clone—but it's a necessary stage of the overall cloning process. Within that specialty, and beyond it, Kraemer has been a pioneer. In the late 1970s, he and colleagues engineered the birth of an addax, a rare African antelope, using artificial insemination with sperm that had been frozen. People at the time asked, Why work with addax? The species, *Addax nasomaculatus,* didn't seem endangered. Now it's extinct everywhere except for a few patches of desert in the southern Sahara. After five decades of quietly working the boundary zone between veterinary medicine and reproductive science, Kraemer is one of the patriarchs in the ART field. Dewey the deer was named in his honor.

For Kraemer, the impetus to work with wildlife came partly from his students, some of whom asked him to teach them skills that might be applied to endangered species. Semen freezing and artificial insemination were proven techniques twenty-five years ago. Embryo transfer and in vitro fertilization showed great promise. Cloning—that was a dream. Kraemer had some small grants to support the student training, but after graduation his young people faced poor odds of landing a job in wildlife or zoo work. "We had told them right up front," he says. "You better have another way of making a living, and you may have to do this on the side." Mark Westhusin, for one, took a job doing research for Granada Bio-Sciences, part of a large cattle company.

Kraemer meanwhile established an effort he called Project Noah's Ark, aimed at putting students and faculty into the field with a mobile laboratory. The lab, in a twenty-eight-foot trailer, was equipped for collecting ova, semen, and tissue samples from threatened populations of wild animals in remote settings, such as the desert bighorn sheep in west Texas. The project's three purposes were to train students, to research techniques, and to preserve frozen tissue samples for possible cloning. Presently the trailer contains a surgical cradle capable of holding an anesthetized bighorn, a portable autoclave (for sterilizing instruments), a laparoscope with fiber optics (for extracting ova from ovaries), three 50-amp generators, and an earnest sign: "ASK NOT ONLY WHAT NATURE CAN DO FOR YOU BUT ALSO WHAT YOU CAN DO FOR NATURE.—D. C. Kraemer." Asking what he could do for Nature by way of assisted reproductive technologies didn't bring Kraemer much financial support. The training has gone forward, but the ark itself is in dry dock.

ANIMAL CLONING BEGAN back in 1951 with frogs. Robert Briggs and Thomas J. King were embryologists, based at a cancer-research institute in Philadelphia, with a medical interest in understanding how genes are turned on and off during embryo development. Briggs, the senior man, figured that a cloning experiment might bring some insight. What he envisioned was transferring the nucleus of a frog cell, taken from an embryo, into an enucleated frog egg—that is, one from which the original nucleus had been scooped out like the pit from an olive. King, hired for his technical skills, would do the micromanipulation, using delicate scissors and tiny glass needles and pipettes. The transferred nucleus would contain a complete set of chromosomes, carrying all the nuclear DNA required for guiding the development of an individual frog. If things went as hoped, the reconfigured egg would divide into two new cells, divide again, and continue dividing through the full course of embryonic growth to yield a living tadpole. From 197 nuclear-transfer attempts, Briggs and King got thirty-five promising embryos, of which twenty-seven survived to the tadpole stage. Although the success rate was low, barely one in eight, their experiment represented a huge triumph. They had proved the principle that an animal could be cloned from a single cell.

Two questions followed. First, could it be done with mammals? Second, could it be done not just from an *embryo* cell, as DNA donor, but from a *mature* cell snipped off an adult? The second question is weighty because cloning from embryo cells is, except under special conditions, cloning blind. If you don't know the adult character of an individual animal—is it healthy, is it beautiful, is it swift, is it meaty, does it have a huge rack of antlers?—why take pains to duplicate it?

For decades, both questions remained in doubt. Nobody succeeded in producing a documented, credible instance of mammal cloning. One researcher claimed to have cloned mice, but his work fell under suspicion, and it could never be verified or repeated. In 1984 two developmental biologists went so far as to state, in the journal *Science*, that their own unsuccessful efforts with mice, as well as other evidence, "suggest that the cloning of mammals by simple nuclear transfer is biologically impossible."

Well, no, it wasn't—as proved, that very year, by a brilliant Danish veterinarian named Steen Willadsen. Unlike the developmental biologists who experimented with frogs or laboratory mice, but like Kraemer and Westhusin, Willadsen focused on farm animals. Working for the British Agricultural Research Council at a laboratory in

Cambridge, he achieved the first verified cloning of a mammal. He did it—a dozen years before the famously cloned bovid, Dolly—with sheep. He took his donor cells from early sheep embryos, which had not yet begun to differentiate into the variously specialized cells (known as somatic cells) that would eventually form body parts. Such undifferentiated cells, it seemed, were crucial. Transferring one nucleus at a time to one enucleated ovum, fusing each pair by means of a gentle electric shock, following that with a few other crafty moves, Willadsen got enough viable embryos to generate three pregnancies, one of which yielded a living lamb. The following year, afloat on his reputation as a cloner, he left Cambridge for Texas, hired away by the same cattle company, Granada, that soon afterward would also hire Mark Westhusin.

"And so," Duane Kraemer says, "Dr. Willadsen then came and taught us how to do cloning."

But Willadsen couldn't teach them to clone an animal from a skin sample sliced off a buck's scrotum, because he hadn't solved the special problems of cloning from somatic cells. Between early embryo cells (which all look alike) and somatic cells (specialized as skin, bone, muscle, nerve, or any sort of internal organ) lurks a deep mystery: the mystery of development and differentiation from a single endowment of DNA. Each cell in a given animal carries a complete copy of the same chromosomal DNA, the same genetic instructions; yet cells respond differently during development, fulfilling different portions of the overall construction plan, assuming different shapes and roles within the body. How does that happen? Why? What tells this cell to become skin, that cell to become bone, another to become liver tissue? What signals them to implement part of the genetic instructions they carry, and to ignore all the rest? Big questions. Cloning researchers, if they were ever to produce an animal cloned from an adult, didn't necessarily need to answer those questions, but they needed to circumvent them. They needed to erase somehow the differentiation of the donor DNA, and to conjure it into operating as though its role within a living creature had just begun anew.

That's what Ian Wilmut, Keith Campbell, and their colleagues in Scotland managed in 1996, using some further touches of biochemical trickery. The result was Dolly, her existence revealed in *Nature* the following year. Dolly's donor cell came from the udder of a six-year-old Finn-Dorset ewe. Her birth was significant because it meant that cloners could now shop before they bought.

LOU HAWTHORNE, a cagey businessman with a trim beard, a weakness for droll language, and a soft heart for animals, tells me how the notion of dog cloning arrived at Texas A&M. Hawthorne is the CEO of a California-based company called Genetic Savings & Clone, which offers the services of "gene banking and cloning of exceptional pets." Another man, Hawthorne's chief financial backer, who prefers to avoid media attention, set the enterprise in motion with a personal whim. "It was just one morning, he was reading the paper," says Hawthorne. "Dolly had been cloned. There was an article about it, and he said: 'I think I'd like to clone Missy. I can afford it.' "

The "he" refers to John Sperling, founder of the Apollo Group, a two-billion-dollar empire that encompasses, among other things, the University of Phoenix, a lucrative enterprise in higher education for working adults. Missy was ten years old at the time of Sperling's brainstorm, a dog of unknown lineage but winning charms, adopted from a pound. Asked by Sperling to make inquiries, Lou Hawthorne solicited proposals from a dozen laboratories; the best came from Texas A&M.

Westhusin remembers telling Hawthorne that they could give it a try but that trying might cost a million dollars a year, take five years, and still be uncertain of success. Okay, said Hawthorne. John Sperling, as he'd declared, could afford it. So the R&D effort toward producing a duplicate Missy—or maybe a multiplicity of copies—began at College Station in 1998. Hawthorne, a word man among scientists, named it the Missyplicity Project.

At the start, it was a joint venture between Texas A&M and an earlier company led by Hawthorne, the Bio-Arts and Research Corporation. Missy contributed a patch of skin cells, which were multiplied by culturing in vitro and then frozen for future use. Westhusin's team gathered a pool of female dogs to serve as egg donors. The eggs, harvested surgically from the oviducts whenever a dog showed signs of ovulation, were emptied of their nuclei using micromanipulation tools (tiny pipettes guided by low-gear control arms within the field of a binocular scope) and then refitted with Missy's DNA by nuclear transfer. These refitted cells were nurtured in the laboratory until some of them showed good embryonic development. Promising embryos were then implanted surgically in ready (that is, estrous) surrogate mothers. Among the factors that make dog cloning difficult is that female canines come into estrus irregularly. Unless you're keeping a riotous kennel, you may not have a bitch in heat when you need her. Westhusin and his Missyplicity partners struggled against that limitation and others for almost five years.

Missy herself died in 2002, still unitary, uncloned. But of course it isn't too late. Her genome is on ice.

Meanwhile, two interesting new entities were born in College Station. One was a cloned cat, the world's first, a little domestic shorthair kitten given the name CC, standing for "copycat." The other was Hawthorne's present company, Genetic Savings & Clone, a for-profit enterprise devoted to the gene-banking of pets (in the form of frozen cells) toward the possibility of their eventual cloning. CC, created with nuclear DNA from a calico donor named Rainbow, was delivered by cesarean section just before Christmas 2001. GSC came into being in response to popular demand.

Alerted to the Missyplicity Project by press reports, dog and cat owners had begun contacting the A&M lab. Some were grieving over recently deceased pets; some were concerned in advance over old animals or sick ones. "We're supposed to be focusing on research here," Westhusin recalls thinking, "and we don't have time to take fifteen phone calls a day and talk to these people about their pets." The callers tended to be emotional, poorly informed, and hopeful. *"I buried him three days ago. Do you think there's any chance if I go dig him up that you could get cells off him?*" Um, I doubt it, Westhusin would say. "*Well, the temperature up here is cold. It's Minnesota . . .*" Westhusin laughs pityingly and so do I. "You want to be nice," he says, "so you sometimes will spend thirty minutes talking on the phone." With the founding of Genetic Savings & Clone, all that grief counseling could be outsourced. Dr. Charles R. Long, another reproductive physiologist and an old friend of Westhusin's, was hired to get the company launched. As general manager, he recruited technical staff and established a research lab to work in partnership with the A&M people. Occasionally he found himself playing psychologist to prospective customers, as Westhusin had done. "The people, the overly emotional ones, many times I would quite frankly try to convince them not to make this decision," Long says. Why? Because they were doing it for the wrong reason. "They were doing it to try to get their special animal back, and you can't get your special animal back. There's no such thing as resurrection. At least not in pets." What you get is just a genetic copy, a new animal with the old DNA, "and it's really important for people to understand that." Chuck Long is a bright, unpretentious man with a small neat mustache, the neck of a linebacker, and huge hands. He once loved a golden retriever named Tex, but he wouldn't have cloned the animal. A loving relationship is about discovery. He'd rather discover a new friend than try to relive life with Tex Two.

Where do people get their misguided ideas about cloning? I ask.

"Hollywood," says Long.

Half ignoring his answer, I press: Do they get them from scientists who oversell the technique or from the media?

"From Hollywood, I think," he repeats. "You know, crazy movies like Arnold Schwarzenegger's *The 6th Day*."

"Was he a clone in that?"

"Yeah, they cloned him."

"I haven't seen that one."

"You've got to see *The 6th Day*. It really stinks."

After a couple of years at Genetic Savings & Clone, Chuck Long parted ways with Lou Hawthorne, and he now works more comfortably for a Texas company, Global Genetics and Biologicals, involved in the production and international export of elite livestock. GSC itself has severed its partnership with Texas A&M and relocated its headquarters to Sausalito, California, with offices overlooking a kayak beach.

Genetic Savings & Clone isn't the only company that sells a gene-banking service for pets; you also might turn to Lazaron BioTechnologies or PerPETuate, Inc. But GSC alone offers the full deal: delivery of clonal duplicates in the near future. The initial cost of putting your pet's genes into the gene bank is $895. Annual storage runs $100. Dogs, with their unique physiological complications (such as opacity of the eggs, making them harder to enucleate), are still problematic. But commercial cat cloning got under way in 2004, with five clients committed, and if all goes well, their cats will be delivered very soon. "In pet cloning," says Hawthorne, "people have an animal that they perceive is extraordinary. In some cases, it's just a perception." In other cases, the extraordinariness is more objective. "You can have an extraordinary mutt," he says. "You can have an animal that has extraordinary intelligence. Extraordinary good looks." Insofar as those traits are genetic, they can be reproduced by cloning, maybe. The delivery price of a healthy young feline, custom-created from DNA of proven appeal, guaranteed to resemble closely your old feline, is fifty thousand dollars. If you think this might meet your emotional expectations, act now.

On the other hand, fifty thousand dollars buys a lot of pretty good cats.

CLONING ENDANGERED SPECIES is a different matter. For starters, who pays? Why does anyone finance this technical approach, seemingly so marginal, rather than putting the money toward basic necessities such as habitat protection? And how can cloning possibly freshen a gene pool that's been reduced to a stagnant puddle? I carry these questions to Dr. Betsy Dresser, director of the Audubon Center for Research of Endangered Species, near New Orleans. ACRES is part of the Audubon Nature Institute, a non-profit group of museums, parks, and other facilities (with no connection to the National Audubon Society). Dresser, who ran a similar research center at the Cincinnati Zoo before coming to New Orleans, has long been a leader in applying captive-breeding efforts and assisted reproductive technologies to endangered species.

She's a brisk, congenial woman, but not easy to get to. ACRES is tucked away in a sunny new building surrounded by bottomland forest at the end of a country road outside the city, on the west bank of the Mississippi River, beneath a towering levee. The land, twelve hundred acres of what once was sugar plantation, is protected by a fence and a guardhouse with an electric gate. A sign says "Freeport McMoRan Audubon Species Survival Center," recognizing the patronship of a mining company in establishing this compound. ACRES itself was created with a fifteen-million-dollar appropriation from the U.S. Fish and Wildlife Service. It resembles the visitor's center of a well-funded state park, but more private. On a morning in April, the air is redolent of honeysuckle. I arrive in time to watch surgery on a domestic cat.

Dr. C. Earle Pope, in a blue smock and mask, is harvesting ova. Several other figures, also in blue, assist him around the operating table. The cat has already been anesthetized and opened, its ovaries exposed. Pope wields a fine forceps in one hand, a hollow steel needle in the other, his head raised to view the target area, which is magnified on a video monitor. He works with easy skill derived from years of experience. The hollow needle is backed by a suction device that feeds into a glass vial on a table nearby. With the forceps, Pope gingerly lifts one ovary so that its follicles (the small, bulbous ovarian sacs) protrude like grapes on a bunch. With the needle, he punctures a follicle and sucks out the egg, then moves to another. The ovary bleeds slightly. Poke, suck, poke, suck, the eggs are whisked away. They accumulate in the vial. When Pope has emptied the follicles of both ovaries, an assistant collects two orange-capped vials and passes them through a window from the operating room to an adjacent lab.

In the lab, which is darkened and barely larger than a closet, a technician moves the eggs from a rinsing solution onto a petri dish. She lifts them, one by one, using an aspirator pipette—that is, with suction applied by her own gentle breath. Her eyes are pressed to a scope. The eggs, surrounded by cloudy globs of ovarian material (called cumulus cells) and air bubbles, are tiny, but they are conspicuous enough to her. You can tell the maturity of ova, she says, by the layers of cumulus cells attached. "These are very good looking." The yield today is twenty-four eggs, about average from a domestic cat. Down the hall, she places the petri dish in an incubator. This afternoon one of Pope's colleagues, Dr. Martha C. Gomez, will enucleate these eggs and endow each with nuclear DNA transferred from an African wildcat.

It won't be the first time such a mix is performed. The African wildcat, *Felis silvestris lybica,* is a tawny little felid native to Africa and the Middle East, closely enough related to the domestic cat, *Felis silvestris catus,* to have figured in earlier experiments involving the two subspecies. Using domestic cat eggs, Gomez, Pope, and their team produced three African wildcat clones in 2003, the eldest born on August 6 and named Ditteaux. (That's *ditto* with Cajun spicing.) The animal from which he and his—his what? not siblings, not twins—two extremely close relatives were cloned, known as Jazz, was itself a product of combined ARTs: the world's first frozen-embryo, thawed-embryo, embryo-transferred wildcat born to a domestic cat. Gomez, Dresser, Pope, and several colleagues coauthored a journal paper on this work, in which they note that the African wildcat "is one of the smallest wild cats, whose future is threatened by hybridization with domestic cats." A person might ask: If hybridization of a wild subspecies with a domestic subspecies is the threat, in what sense is mixing nuclei from one subspecies with eggs from the other subspecies a solution?

Another skeptical question, which I put to Betsy Dresser, is whether this fancy stuff can somehow mitigate the problem of low genetic diversity in sorely endangered species. If it can't, what's the point? "Well, indeed it can," she says. "What we're trying to do is use cloning to bring in the genetic material from animals that are not reproducing." Among any population, she notes, there are always infertile individuals, marginalized individuals, elderly or unlucky individuals, who fail to breed and so contribute no genes to the next generation. In a large population (though Dresser doesn't mention this point), their exclusion represents Darwinian selection, which drives evolution. But in a very small population (she notes rightly), their participation could be crucial. "If you can use the genetic material from those individuals, it helps widen the genetic pool a bit."

Imagine you've got a captive population of just five black-footed ferrets, with no others surviving on the planet. Four of your ferrets are males, and the fifth is a post-reproductive female. One young male chokes to death while eating a prairie dog with reckless gusto. What do you do? Of course you grab the old female and the dead male, take tissue samples, and clone them. But wait—in this scenario of five, there are no viable black-footed ferret ova to receive the clonal DNA. So you use the next best thing: enucleated eggs from a mink. Then you breed your cloned female with one of the males, breed any daughters she produces with other males, get the cloned male's genes into the reproductive jumble too, and thereby postpone (maybe indefinitely) the doom of your miserable little population. Whether your ferrets ever go back into the wild is another question. Do you dare send them? Do you keep breeding and cloning until you've got a few dozen, a few hundred? All this would be expensive at best and, if you hadn't meanwhile solved the root causes of endangerment (such as insufficient habitat, poaching, or exotic species inflicting too much predation or competition), ultimately futile. No clones of an endangered species, and no descendants of clones, have yet been released to the wild.*

*It bears noting, though, that other ART methods have yielded some returns to the wild, such as the hundred Mississippi sandhill cranes, additions to an endangered subspecies, that were bred at ACRES using artificial insemination and released on a refuge near Pascagoula.

What about the money issue? I ask Dresser. Are resources being diverted that might otherwise pay for habitat preservation? Her answer is candid: "The money that comes to this kind of research is primarily from people that are not going to support habitat." She's a skilled fund-raiser as well as a respected scientist; she has been through this in Cincinnati, now New Orleans, and she knows her constituency. Supporting the research arm of a fine metropolitan zoo is a bit like supporting the symphony, the conservatory, the opera. These people "don't want to give their money to Africa, or to Asia, or somewhere. They don't want their money in political environments where they're never going to see their name on a plaque." At the various branches of the Audubon Nature Institute, including ACRES, there are more than a few grateful plaques.

Back on the city side of the river, I visit the Audubon Zoo on Magazine Street for a glimpse of Ditteaux the cloned wildcat, temporarily on display there. For this interlude of public exposure, he lives in a glass-fronted cage furnished with small boulders, trees, and a six-foot square of scenery meant to approximate northern Africa. He's a handsome animal, lanky and lithe, nervous, his brownish-gray fur marked with pale stripes. As I watch, his pale green eyes come alert to something—the sight of a squirrel outside the building, visible through an opposite window.

Groups of schoolchildren pass Ditteaux's cage. A well-fed boy in an orange T-shirt reads the sign and then asks, "It's a clone?" Yes. With some vehemence, he says, "Okay, that's *freaky*."

WHATEVER THE DOWNSIDE OF INVESTING MONEY and time in such an approach to endangered species, at least one private firm has also done it: Advanced Cell Technology, of Worcester, Massachusetts. Founded originally as a subsidiary of a poultry genetics business, ACT now concentrates mainly on human medical issues. The company's work with wildlife is an adventuresome sideline, bearing no such commercial promise as cloning whitetail deer for the trophy market but offering the possibility of a public good, roughly equivalent to pro bono work by a law firm. It also offers, when successful, good publicity.

In early 2001, ACT announced that "the first cloned endangered animal," an eighty-pound male gaur, had been born to a surrogate mother. The gaur is a species of wild cattle, *Bos gaurus,* native to southeastern Asia from Thailand to Nepal. Calling it an "endangered animal" was mildly misleading; the international body that keeps track of such things classifies the gaur as "Vulnerable," not actually "Endangered," with somewhere between 13,000 and 30,000 individuals in the wild. But the population is declining, and the trend isn't likely to reverse. Vulnerable or endangered, the species deserves attention.

Two technical points made ACT's gaur work especially notable. First, the nuclear DNA came from gaur cells, derived from a tissue sample that had sat frozen for eight years in a gene bank at the San Diego Zoo. Second, the enucleated egg cell into which the gaur DNA had been transferred came from a domestic cow. So this too was a case of cross-species cloning—in fact, it was the first recorded case, a precursor to the African wildcat project in New Orleans. Arguably, the technique could be valuable in situations when egg cells of an endangered species are unavailable—when there are no surviving females, say, or so few that you wouldn't dare cut one open to harvest her eggs.

What made the case less encouraging was that the baby gaur, named Noah, died of dysentery within two days. Its death fell hard on Robert P. Lanza, a vice president of ACT, who had led the cloning effort.

At that time, Lanza had nearly sealed an agreement with Spanish officials toward cloning an extinct Spanish subspecies of mountain goat, the bucardo. The bucardo *(Capra pyrenaica pyrenaica)* had languished at desperately low population levels throughout the twentieth century, probably because of competition with livestock, diseases caught from livestock, poaching, and other travails. The last one died in 2000, clunked by a falling tree, but provident biologists had arranged to freeze some of its tissue for posterity. Lanza hoped to clone the bucardo back into existence, using the frozen sample for nuclear DNA, a domestic nanny goat as egg donor, and a nanny again as surrogate to carry the fetus. That plan collapsed with the death of Noah the gaur. Two years later ACT's cloning team tried again, this time achieving the birth of two cloned calves from another species of wild Asian cattle, the banteng, *Bos javanicus.* The banteng is unambiguously endangered, with no more than eight thousand individuals in the wild. The nuclear DNA came from another frozen sample that had been stored, for twenty-five years, at the San Diego Zoo.

The gene bank in San Diego, loosely known as the Frozen Zoo, was established three decades ago by a pathologist named Kirk Benirschke, who was soon joined by a young geneticist, Oliver A. Ryder. Benirschke and Ryder foresaw that these cell samples might be useful in genetic studies of relatedness among wild species. They didn't foresee that the frozen cells might be cloned back to life. The collection now represents about 7,000 individual animals of 450 different species; about half of those samples came from animals resident at the San Diego Zoo, the rest from animals at other zoos and captive facilities, or from the wild. Ryder is still there, the man to see if you want a morsel of rare or endangered DNA for some legitimate purpose. Cloners across the country, from College Station to Worcester and beyond, point to San Diego's Frozen Zoo as a prescient enterprise that should be emulated widely, preserving as much genetic diversity as possible from endangered species before their populations decline too far. Ryder, for his part, supports the idea of cloning when it might return a valuable genotype to a breeding population. The original banteng whose frozen cells went to ACT, for instance, died in 1980 without offspring, having made no genetic contribution to the captive banteng population. One of the two clones produced from those cells was healthy, and that animal has since been returned to San Diego; its lost genes may eventually be bred back into the zoo population of banteng, possibly adding some much needed diversity.

But gene banking is no panacea. Ryder himself says: "I think it's gonna be a somber day when we realize that the only thing left of a species is something we've got in the Frozen Zoo."

AMONG EXTINCT SPECIES AND SUBSPECIES, the bucardo goat represents a good prospect for cloning, because the extinction is so recent and the cell sample was properly preserved. Less propitious circumstances, though, don't prevent people from trying to resurrect a lost beast.

Scientists at Kinki University in Japan have begun work toward cloning a woolly mammoth, using tissue samples from a twenty-thousand-year-old carcass recently excavated from frozen Siberian tundra. Elephants, the mammoth's closest living relatives, will serve as egg donors and surrogate mothers, if the project ever gets that far. Cloning researchers at the Australian Museum in Sydney hope to re-create the thylacine, a carnivorous marsupial loosely known as the Tasmanian tiger, last seen alive in 1936. For that effort, the starting point is a thylacine pup stored in alcohol since 1866. Alcohol is a gentler preservative than formaldehyde, and the Australians have managed to extract some DNA fragments in fairly good condition—but no complete DNA strands, let alone any viable thylacine cells with nuclei that could be transferred intact. The optimistic Aussies aim to reassemble their squibs and scraps into a full set of thylacine DNA, perhaps patching the gaps with genetic material from other marsupials. Plausible? Not very, according to Ryder. "What's the chance that you could shred the phone book," he asks, "and then drop it out of a window and have it come back together?" Once they have reassembled their phone book, if they do, the Australians will create artificial chromosomes for insertion into an egg from some related species, such as the Tasmanian devil. Meanwhile in Hyderabad, India, a team led by Dr. Lalji Singh proposes to clone an Asiatic cheetah, a subspecies extinct in India for the past fifty years. They want to use nuclear DNA from a cheetah loaned by Iran, though Iran itself has only a few dozen cheetahs in the wild, and none of those has been promised so far. If the Indians do get their chance to proceed, the eggs and the surrogate wombs will be furnished by leopards.

Each of these projects, variously dreamy or doable, represents an effort at cross-species cloning, like the banteng-and-cow work by ACT. This sort of trick raises further issues. What are the physiological consequences of mixing nuclear DNA from a cheetah with mitochondrial DNA (which comes along with the enucleated egg and helps regulate the cell's biochemistry) from a leopard? What are the ecological implications of mixing mammoths with elephants in a world where the mammoths' ecosystem no longer exists? What's the merit or demerit of blurring lines between species (cheetah and leopard, thylacine and devil) by means of laboratory gimmickry, in order to "preserve" a vanishing subspecies or "restore" an extinct species in the wild?

Lines, their integrity or transgression, are exactly what's at issue: the line between one species and another that defines biological diversity, the line between one animal and another that constitutes individuality, the line between living and dead that gives meaning—as well as poignant temporal limit—to life. And yet those lines aren't always easy to draw, let alone to enforce or respect. Even species, even in the wild, sometimes blur into one another: wolves breeding with coyotes, blue-winged warblers with gold-winged warblers, barn swallows with house martins, mule deer with whitetails. True, these natural mongrelizations represent exceptions to the rule of how species are generally demarcated. But they complicate any effort to think clearly about drawing other lines, such as the line between *Felis silvestris lybica* and *Felis silvestris catus,* the line between embryo transfer and nuclear transfer, the line between genetically modified organisms and heirloom tomatoes (which have themselves been genetically modified by generations of careful horticulture), the line between extinct and merely frozen, the line between what we can do and what we should do, the line between nature and ART.

Recognizing such complications is not necessarily the same as surrendering to a paralyzing relativism. Lines that suggest boundaries of ethical behavior, of judicious balance between opposing concerns, and of precious entities deserving preservation are important even when they reveal themselves, at close inspection, to be smeary zones of gradated gray. The mapping of such boundaries can't be done by science, which is capable of measuring shades of gray but not choosing among them. That leaves religion, philosophy, social consensus, and common sense. Which of those do we rely on for decisions about assisted reproductive technologies, such as cloning, when the species being assisted is not the banteng or the whitetail deer but *Homo sapiens*?

Consider the prospect of germline genetic engineering—that is, fiddling with genes in human embryo cells before those cells are grown into human fetuses. Germline engineering is not yet available as a consumer option, for medical purposes or any others, but soon it may be. Select genes would be added to, subtracted from, or modified in an embryo cell, after which the cell would be cloned into a customized human child. This process would permit the correction of genetic weaknesses—bad eyesight, for instance, or sickle-cell anemia—in advance of birth. When that starts happening, as Bill McKibben has warned in his book *Enough: Staying Human in an Engineered Age*, "the line between fixing problems and 'enhancing' offspring" will disappear, at least for any parents who want their kids to be as bright, robust, good-looking, and competitive as humanly (that is, technologically) possible. If you can repair your future child's myopia with preemptive genetic tinkering, you might also want to increase her IQ by a few dozen points. Will it lead to a world as utopian as Lake Wobegon, where all the children are above average? Of course not. It will just add genetic manipulation of embryos and child cloning to the means by which affluent, fussy people try to distance themselves from bad luck, disappointment, menial work, death, and poor people.

McKibben, his ardent humaneness informed by a lot of careful research and thinking, proposes that we should recoil from such possibilities and declare "Enough!" He suggests that somewhere amid the dizzying possibilities of ART as applied to humans, beyond fertility medicine but short of germline genetic engineering, we might locate "the enough line"—that is, the threshold of ugly and corruptive weirdness across which a wholesome person and a wise society do not go.

As much as I want to agree with him, my own survey of animal cloning forces me to conclude that his "enough" line, like any I might try to draw myself, is as subjective as it is sensible. There is, in fact, no line. There is only a spectrum: a set of choices among shades of gray. Of course, that's not to say some choices aren't nuttier than others.

Cloning adult humans, for instance. Any thorough discussion of assisted reproductive technologies comes eventually to this topic, which the animal-cloning scientists detest and dismiss but which other people consider central. The animal guys are right—it's not central—but, like a parrot in a cage of canaries, it's too big and noisy to ignore. What if John Sperling or some other loopy billionaire decides one morning to commission not the cloning of his lovable mutt but the cloning of himself? If that decision hasn't already been made, quietly in a penthouse somewhere, it probably soon will be; and whatever unique technical difficulties or scientific scruples have so far prevented the consummation of such a desire will soon be overcome. Some people view the prospect of human cloning with great alarm. Bill Clinton labeled it "morally reprehensible." His presidential ethics commission recommended federal laws to prohibit human cloning. Finding myself less certain than Clinton or those advisers about the moral or legal verities against which human cloning should be measured, I'd simply call it perniciously stupid. Then again, many things people do nowadays are, in my opinion, perniciously stupid. Not all of them are illegal, and so, I suppose, human cloning needn't be either.

Down in College Station, I'm reminded of all this during my chat with Duane Kraemer, when we bounce from the subject of endangered species back to companion animals. Isn't there something misguided, I ask Kraemer, about cloning your pet? Doesn't it reflect an inclination to deny mortality?

Deny mortality? "We do that every day!" he says brightly. "We get up and brush our teeth. Why do we do that? Because we want to live as long as we can. So denial of mortality is, yeah, it's in our being. And it's not only natural. It's necessary."

TWO OTHER VOICES OF WISDOM ECHO through my head, addressing aspects of the question why. One of these voices belonged to J. Robert Oppenheimer, the physicist and founding director of the Los Alamos nuclear-weapons laboratory. Trust me on this seeming digression. Having helped build the first atomic bomb, Oppenheimer resisted the notion that America should rush ahead to build a thermonuclear superbomb. It was fission versus fusion, uranium versus hydrogen, kilotons versus megatons, and the global political context of 1943 versus the context of 1951. His resistance was swept aside by a clever design principle concocted by two other physicists, one of whom was Edward Teller. Asked later by an inquisitorial panel about how the H-bomb decision was made, Oppenheimer declined to speak about technical details. "However," he said mordantly, "it is my judgment in these things that when you see something that is technically sweet, you go ahead and do it and you argue about what to do about it only after you have had your technical success." This scary truth, which might be thought of as Oppenheimer's Axiom, explains many controversial gambits in whizbang scientific engineering. Why do some scientists crave to clone animals? Not just because they can but because they can do so—with an elegant medley of ingenious laboratory moves—in a way that is technically sweet. And therefore irresistible.

The other voice comes from Louis Armstrong, as recorded in 1931:

> Oh, when skies are cloudy and gray,
> They're only gray for a day,
> Bay-bay-bay-bee . . .
> So wrap your troubles in dreams,
> And dream your troubles away.

Duane Kraemer is right in noting that this methodology isn't unique to assisted reproductive technologists.

On the morning after our conversation, Dr. Kraemer welcomes me to his home, in a neighborhood just north of the A&M campus, to meet the famous cloned house cat, CC. As we enter, she crosses a living room of draped-over furniture and leaps onto a carpeted cat perch, presenting herself for Kraemer's gentle petting. She's no longer a kitten. She arches her back to my touch, then carefully sniffs my hand. Her fur is soft and clean. She looks like any normal cat. The most striking aspect of her appearance, which I wouldn't notice if I hadn't read some background, is that she's a tiger-tabby shorthair, mottled black-and-gray with a white breast and legs. It's striking because she was cloned from a calico.

CC's color pattern is utterly different from that of Rainbow, her DNA donor. The cause of this difference is complicated (involving random inactivation of one of her two X chromosomes, which in a female such as CC are redundantly paired, though each may carry a distinct gene for color), but it can be simplified in a single word: random. Cloning isn't resurrection, as the man said. It's not even, quite, duplication.

On CC's right cheek, otherwise white, I notice a small patch of tan fur, like a birthmark. Yes, says Kraemer, that wasn't present in Rainbow either. The genotype may be identical in a clone, but it gets expressed differently. Maybe one day when she was a fetus, inside the surrogate mother, CC rubbed her little face against the wall of the womb. A smudge. Things happen.

CHARLES C. MANN

The Coming Death Shortage

FROM THE *ATLANTIC MONTHLY*

The ever-lengthening human lifespan may seem like the goal of medical science, but it carries with it some disturbing consequences. Charles C. Mann contemplates the dislocations and intergenerational warfare that increased longevity will bring.

Anna Nicole Smith's role as a harbinger of the future is not widely acknowledged. Born Vickie Lynn Hogan, Smith first came to the attention of the American public in 1993, when she earned the title Playmate of the Year. In 1994 she married J. Howard Marshall, a Houston oil magnate said to be worth more than half a billion dollars. He was eighty-nine and wheelchair-bound; she was twenty-six and quiveringly mobile. Fourteen months later Marshall died. At his funeral the widow appeared in a white dress with a vertical neckline. She also claimed that Marshall had promised half his fortune to her. The inevitable litigation sprawled from Texas to California and occupied batteries of lawyers, consultants, and public relations specialists for more than seven years.

Even before Smith appeared, Marshall had disinherited his older son. And he had infuriated his younger son by lavishing millions on a mistress, an exotic dancer, who then died in a bizarre face-lift accident. To block Marshall senior from squandering on Smith money that Marshall junior regarded as rightfully his, the son seized control of his father's assets by means that the trial judge later said were so "egregious," "malicious," and "fraudulent" that he regretted being unable to fine the younger Marshall more than forty-four million dollars in punitive damages.

In its epic tawdriness the Marshall affair was natural fodder for the tabloid media. Yet one aspect of it may soon seem less a freak show than a cliché. If an increasingly influential group of researchers is correct, the lurid spectacle of intergenerational warfare will become a typical social malady.

The scientists' argument is circuitous but not complex. In the past century U.S. life expectancy has climbed from forty-seven to seventy-seven, increasing by nearly two thirds. Similar rises happened in almost every country. And this process shows no sign of stopping: according to the United Nations, by 2050 global life expectancy will have increased by another ten years. Note, however, that this tremendous increase has been in *average* life expectancy—that is, the number of years that most people live. There has been next to no increase in the *maximum* lifespan, the number of years that one can possibly walk the earth—now thought to be about 120. In the scientists' projections, the ongoing increase in average lifespan is about to be joined by something never before seen in human history: a rise in the maximum possible age at death.

Stem cell banks, telomerase amplifiers, somatic gene therapy—the list of potential longevity treatments incubating in laboratories is startling. Three years ago a multi-institutional scientific team led by Aubrey de Grey, a theoretical geneticist at Cambridge University, argued in a widely noted paper that the first steps toward "engineered negligible senescence"—a rough-and-ready version of immortality—would have "a good chance of success in mice within ten years." The same techniques, De Grey says, should be ready for human beings a decade or so later. "In ten years we'll have a pill that will give you twenty years," says Leonard Guarente, a professor of biology at MIT. "And then there'll be another pill after that. The first hundred-and-fifty-year-old may have already been born."

Critics regard such claims as wildly premature. In March ten respected researchers predicted in the *New England Journal of Medicine* that "the steady rise in life expectancy during the past two centuries may soon come to an end," because rising levels of obesity are making people sicker. The research team leader, S. Jay Olshansky, of the University of Illinois School of Public Health, also worries about the "potential impact of infectious disease." He believes that medicine can and will overcome these problems; his "cautious and I think defensibly optimistic estimate" is that the average lifespan will reach eighty-five or ninety—in 2100. Even this relatively slow rate of increase, he says, will radically alter the underpinnings of human existence. "Pushing the outer limits of lifespan" will force the world to confront a situation no society has ever faced before: an acute shortage of dead people.

The twentieth-century jump in life expectancy transformed society. Fifty years ago senior citizens were not a force in electoral politics. Now the AARP is widely said to be the most powerful organization in Washington. Medicare, Social Security, retirement, Alzheimer's, snowbird economies, the population boom, the golfing boom, the cosmetic-surgery boom, the nostalgia boom, the recreational-vehicle boom, Viagra—increasing longevity is entangled in every one. Momentous as these changes have been, though, they will pale before what is coming next.

From religion to real estate, from pensions to parent-child dynamics, almost every aspect of society is based on the orderly succession of generations. Every quarter century or so children take over from their parents—a transition as fundamental to human existence as the rotation of the planet about its axis. In tomorrow's world, if the optimists are correct, grandparents will have living grandparents; children born decades from now will ignore advice from people who watched the Beatles on *The Ed Sullivan Show.* Intergenerational warfare—the Anna Nicole Smith syndrome—will be but one consequence. Trying to envision such a world, sober social scientists find themselves discussing pregnant seventy-year-olds, offshore organ farms, protracted adolescence, and lifestyles policed by insurance companies. Indeed, if the biologists are right, the coming army of centenarians will be marching into a future so unutterably different that they may well feel nostalgia for the long-ago days of three score and ten.

The oldest *in vitro* fertilization clinic in China is located on the sixth floor of a no-star hotel in Changsha, a gritty flyover city in the south-central portion of the country. It is here that the clinic's founder and director, Lu Guangxiu, pursues her research into embryonic stem cells.

Most cells *don't* divide, whatever elementary school students learn—they just get old and die. The body subcontracts out the job of replacing them to a special class of cells called stem cells. Embryonic stem cells—those in an early-stage embryo—can grow into any kind of cell: spleen, nerve, bone, whatever. Rather than having to wait for a heart transplant, medical researchers believe, a patient could use stem cells to grow a new heart: organ transplant without an organ donor.

The process of extracting stem cells destroys an early-stage embryo, which has led the Bush administration to place so many strictures on stem cell research that scientists complain it has been effectively banned in this country. A visit to Lu's clinic not long ago suggested that ultimately Bush's rules won't stop anything. Capitalism won't let them.

During a conversation Lu accidentally brushed some papers to the floor. They were faxes from venture capitalists in San Francisco, Hong Kong, and Stuttgart. "I get those all the time," she said. Her operation was short of money—a chronic problem for scientists in poor countries. But it had something of value: thousands of frozen embryos, an inevitable by-product of *in vitro* fertilizations. After obtaining permission from patients, Lu uses the embryos in her work. It is possible that she has access to more embryonic stem cells than all U.S. researchers combined.

Sooner or later, in one nation or another, someone like Lu will cut a deal: frozen embryos for financial backing. Few are the stem-cell researchers who believe that their work will not lead to tissue-and-organ farms, and that these will not have a dramatic impact on the human lifespan. If Organs 'Я' Us is banned in the United States, Americans will fly to longevity centers elsewhere. As Steve Hall wrote in *Merchants of Immortality,* biotechnology increasingly resembles the software industry. Dependence on venture capital, loathing of regulation, pathological secretiveness, penchant for hype, willingness to work overseas—they're all there. Already the U.S. Patent Office has issued four hundred patents concerning human stem cells.

Longevity treatments will almost certainly drive up medical costs, says Dana Goldman, the director of health economics at the RAND Corporation, and some might drive them up significantly. Implanted defibrillators, for example, could constantly monitor people's hearts for signs of trouble, electrically regulating the organs when they miss a beat. Researchers believe that the devices would reduce heart-disease deaths significantly. At the same time, Goldman says, they would by themselves drive up the nation's health-care costs by "many billions of dollars" (Goldman and his colleagues are working on nailing down how much), and they would be only one of many new medical interventions. In developed nations anti-retroviral drugs for AIDS typically cost about fifteen thousand dollars a year. According to James Lubitz, the acting chief of the aging and chronic-disease statistics branch of the CDC National Center for Health Statistics, there is no a priori reason to suppose that lifespan extension will be cheaper, that the treatments will have to be administered less frequently, or that their inventors will agree to be less well compensated. To be sure, as Ramez Naam points out in *More Than Human,* which surveys the prospects for "biological enhancement," drugs inevitably fall in price as their patents expire. But the same does not necessarily hold true for medical procedures: heart bypass operations are still costly, decades after their invention. And in any case there will invariably be newer, more effective, and more costly drugs. Simple arithmetic shows that if eighty million U.S. senior citizens were to receive fifteen thousand dollars' worth of treatment every year, the annual cost to the nation would be $1.2 trillion—"the kind of number," Lubitz says, "that gets people's attention."

The potential costs are enormous, but the United States is a rich nation. As a share of gross domestic product the cost of U.S. health care roughly doubled from 1980 to the present, explains David M. Cutler, a health-care economist at Harvard. Yet unlike many cost increases, this one signifies that people are better off. "Would you rather have a heart attack with 1980 medicine at the 1980 price?" Cutler asks. "We get more and better treatments now, and we pay more for the additional services. I don't look at that and see an obvious disaster."

The critical issue, in Goldman's view, will be not the costs per se but determining who will pay them. "We're going to have a very public debate about whether this will be covered by insurance," he says. "My sense is that it won't. It'll be like cosmetic surgery—you pay out of pocket." Necessarily, a pay-as-you-go policy would limit access to longevity treatments. If high-level anti-aging therapy was expensive enough, it could become a perk for movie stars, politicians, and CEOs. One can envision Michael Moore fifty years from now, still denouncing the rich in political tracts delivered through the next generation's version of the Internet—neural implants, perhaps. Donald Trump, a one-hundred-eight-year-old multibillionaire in 2054, will be firing the children of the apprentices he fired in 2004. Meanwhile, the maids, chauffeurs, and gofers of the rich will stare mortality in the face.

Short of overtly confiscating rich people's assets, it would be hard to avoid this divide. Yet as Goldman says, there will be "furious" political pressure to avert the worst inequities. For instance, government might mandate that insurance cover longevity treatments. In fact, it is hard to imagine any democratic government foolhardy enough *not* to guarantee access to those treatments, especially when the old are increasing in number and political clout. But forcing insurers to cover longevity treatments would only change the shape of the social problem. "Most everyone will want to take [the treatment]," Goldman says. "So that jacks up the price of insurance, which leads to more people uninsured. Either way, we may be bifurcating society."

Ultimately, Goldman suggests, the government would probably end up paying outright for longevity treatments: an enormous new entitlement program. How could it be otherwise? Older voters would want it because it is in their interest; younger ones would want it because they, too, will age. "At the same time," he says, "nobody likes paying taxes, so there would be constant pressure to contain costs."

To control spending, the program might give priority to people with healthy habits; no point in retooling the genomes of smokers, risk takers, and addicts of all kinds. A kind of reverse eugenics might occur, in which governments would freely allow the birth of people with "bad" genes but would let nature take its course on them as they aged. Having shed the baggage of depression, addiction, mental retardation, and chemical-sensitivity syndrome, tomorrow's legions of perduring old would be healthier than the young. In this scenario moralists and reformers would have a field day.

Meanwhile, the gerontocratic elite will have a supreme weapon against the young: compound interest. According to a 2004 study by three researchers at the London Business School, historically the average rate of real return on stock markets worldwide has been about five percent. Thus a twenty-year-old who puts $10,000 in the market in 2010 should expect by 2030 to have about $27,000 in real terms—a tidy increase. But that happy forty-year-old will be in the same world as septuagenarians and octogenarians who began investing their money during the Carter administration. If someone who turned seventy in 2010 had invested $10,000 when he was twenty, he would have about $115,000. In the same twenty-year period during which the young person's account grew from $10,000 to $27,000, the old person's account would grow from $115,000 to $305,000. Inexorably, the gap between them will widen.

The result would be a tripartite society: the very old and very rich on top, beta-testing each new treatment on themselves; a mass of the ordinary old, forced by insurance into supremely healthy habits, kept alive by medical entitlement; and the diminishingly influential young. In his novel *Holy Fire* (1996) the science-fiction writer and futurist Bruce Sterling conjured up a version of this dictatorship-by-actuary: a society in which the cautious, careful centenarian rulers, supremely fit and disproportionately affluent if a little frail, look down with ennui and mild contempt on their juniors. Marxist class warfare, upgraded to the biotech era!

In the past, twenty- and thirty-year-olds had the chance of sudden windfalls in the form of inheritances. Some economists believe that bequests from previous generations have provided as much as a quarter of the start-up capital for each new one—money for college tuitions, new houses, new businesses. But the image of an ingénue's getting a leg up through a sudden bequest from Aunt Tilly will soon be a relic of late-millennium romances.

Instead of helping their juniors begin careers and families, tomorrow's rich oldsters will be expending their disposable income to enhance their memories, senses, and immune systems. Refashioning their flesh to ever higher levels of performance, they will adjust their metabolisms on computers, install artificial organs that synthesize smart drugs, and swallow genetically tailored bacteria and viruses that clean out arteries, fine-tune neurons, and repair broken genes. Should one be reminded of H. G. Wells's *The Time Machine*, in which humankind is divided into two species, the ethereal Eloi and the brutish, underground-dwelling Morlocks? "As I recall," Goldman told me recently, "in that book it didn't work out very well for the Eloi."

WHEN LIFESPANS EXTEND INDEFINITELY, the effects are felt throughout the life cycle, but the biggest social impact may be on the young. According to Joshua Goldstein, a demographer at Princeton, adolescence will in the future evolve into a period of experimentation and education that will last from the teenage years into the mid-thirties. In a kind of *wanderjahr* prolonged for decades, young people will try out jobs on a temporary basis, float in and out of their parents' homes, hit the Europass-and-hostel circuit, pick up extra courses and degrees, and live with different people in different places. In the past the transition from youth to adulthood usually followed an orderly sequence: education, entry into the labor force, marriage, and parenthood. For tomorrow's thirtysomethings, suspended in what Goldstein calls "quasi-adulthood," these steps may occur in any order.

From our short-life-expectancy point of view, quasi-adulthood may seem like a period of socially mandated fecklessness—what Leon Kass, the chair of the President's Council on Bioethics, has decried as the coming culture of "protracted youthfulness, hedonism, and sexual license." In Japan, ever in the demographic forefront, as many as one out of three young adults is either unemployed or working part-time, and many are living rent-free with their parents. Masahiro Yamada, a sociologist at Tokyo Gakugei University, has sarcastically dubbed them *parasaito shinguru*, or "parasite singles." Adult offspring who live with their parents are common in aging Europe, too. In 2003 a report from the British Prudential financial-services group awarded the 6.8 million British in this category the mocking name of "kippers"—"kids in parents' pockets eroding retirement savings."

To Kass, the main cause of this stasis is "the successful pursuit of longer life and better health." Kass's fulminations easily lend themselves to ridicule. Nonetheless, he is in many ways correct. According to Yuji Genda, an economist at Tokyo University, the drifty lives of parasite singles are indeed a by-product of increased longevity, mainly because longer-lived seniors are holding on to their jobs. Japan, with the world's oldest population, has the highest percentage of working senior citizens of any developed nation: one out of three men over sixty-five is still on the job. Everyone in the nation, Genda says, is "tacitly aware" that the old are "blocking the door."

In a world of two-hundred-year-olds "the rate of rise in income and status perhaps for the first hundred years of life will be almost negligible," the crusty maverick economist Kenneth Boulding argued in a prescient article from 1965. "It is the propensity of the old, rich, and powerful to die that gives the young, poor, and powerless hope." (Boulding died in 1993, opening up a position for another crusty maverick economist.)

Kass believes that "human beings, once they have attained the burdensome knowledge of good and bad, should not have access to the tree of life." Accordingly, he has proposed a straightforward way to prevent the problems of youth in a society dominated by the old: "resist the siren song of the conquest of aging and death." Senior citizens, in other words, should let nature take its course once humankind's biblical seventy-year lifespan is up. Unfortunately, this solution is self-canceling, since everyone who agrees with it is eventually eliminated. Opponents, meanwhile, live on and on. Kass, who is sixty-six, has another four years to make his case.

Increased longevity may add to marital strains. The historian Lawrence Stone was among the first to note that divorce was rare in previous centuries partly because people died so young that bad unions were often dissolved by early funerals. As people lived longer, Stone argued, divorce became "a functional substitute for death." Indeed, marriages dissolved at about the same rate in 1860 as in 1960, except that in the nineteenth century the dissolution was more often due to the death of a partner, and in the twentieth century to divorce. The corollary that children were as likely to live in households without both biological parents in 1860 as in 1960 is also true. Longer lifespans are far from the only reason for today's higher divorce rates, but the evidence seems clear that they play a role. The prospect of spending another twenty years sitting across the breakfast table from a spouse whose charm has faded must have already driven millions to divorce lawyers. Adding an extra decade or two can only exacerbate the strain.

Worse, child-rearing, a primary marital activity, will be even more difficult than it is now. For the past three decades, according to Ben J. Wattenberg, a senior fellow at the American Enterprise Institute, birth rates around the world have fallen sharply as women have taken advantage of increased opportunities for education and work outside the home. "More education, more work, lower fertility," he says. The title of Wattenberg's latest book, published in October, sums up his view of tomorrow's demographic prospects: *Fewer.* In his analysis, women's continuing movement outside the home will lead to a devastating population crash—the mirror image of the population boom that shaped so much of the past century. Increased longevity will only add to the downward pressure on birth rates, by making childbearing even more difficult. During their twenties, as Goldstein's quasi-adults, men and women will be unmarried and relatively poor. In their thirties and forties they will finally grow old enough to begin meaningful careers—the worst time to have children. Waiting still longer will mean entering the maelstrom of reproductive technology, which seems likely to remain expensive, alienating, and prone to complications. Thus the parental paradox: increased longevity means *less* time for pregnancy and child-rearing, not more.

Even when women manage to fit pregnancy into their careers, they will spend a smaller fraction of their lives raising children than ever before. In the mid-nineteenth century white women in the United States had a life expectancy of about forty years and typically bore five or six children. (I specify Caucasians because records were not kept for African Americans.) These women literally spent more than half their lives caring for offspring. Today U.S. white women have a life expectancy of nearly eighty and bear an average of 1.9 children—below replacement level. If a woman spaces two births close together, she may spend only a quarter of her days in the company of offspring under the age of eighteen. Children will become ever briefer parentheses in long, crowded adult existences. It seems inevitable that the bonds between generations will fray.

Purely from a financial standpoint, parenthood has always been a terrible deal. Mom and Dad fed, clothed, housed, and educated the kids but received little in the way of tangible return. Ever since humankind began acquiring property, wealth has flowed from older generations to younger ones. Even in those societies where children herded cattle and tilled the land for their aged progenitors, the older generation consumed so little and died off so quickly that the net movement of assets and services was always downward. "Of all the misconceptions that should be banished from discussions of aging," F. Landis MacKellar, an economist at the International Institute for Applied Systems Analysis, in Austria, wrote in the journal *Population and Development Review* in 2001, "the most persistent and egregious is that in some simpler and more virtuous age children supported their parents."

This ancient pattern changed at the beginning of the twentieth century, when government pension and social-security schemes spread across Europe and into the Americas. Within the family, parents still gave much more than they received, according to MacKellar, but under the new state plans the children in effect banded together outside the family and collectively reimbursed the parents. In the United States workers pay less to Social Security than they eventually receive; retirees are subsidized by the contributions of younger workers. But on the broadest level financial support from the young is still offset by the movement of assets within families—a point rarely noted by critics of "greedy geezers."

Increased longevity will break up this relatively equitable arrangement. Here concerns focus less on the super-rich than on middle-class senior citizens, those who aren't surfing the crest of compound interest. These people will face a Hobson's choice. On the one hand, they will be unable to retire at sixty-five because the young would end up bankrupting themselves to support them—a reason why many would-be reformers propose raising the retirement age. On the other hand, it will not be feasible for most of tomorrow's nonagenarians and centenarians to stay at their desks, no matter how fit and healthy they are.

The case against early retirement is well known. In economic jargon the ratio of retirees to workers is known as the "dependency ratio," because through pension and Social Security payments people who are now in the work force funnel money to people who have left it. A widely cited analysis by three economists at the Organization for Economic Cooperation and Development estimated that in 2000 the overall dependency ratio in the United States was 21.7 retirees for every 100 workers, meaning (roughly speaking) that everyone older than sixty-five had five younger workers contributing to his pension. By 2050 the dependency ratio will have almost doubled, to 38 per 100; that is, each retiree will be supported by slightly more than two current workers. If old-age benefits stay the same, in other words, the burden on younger workers, usually in the form of taxes, will more than double.

This may be an underestimate. The OECD analysis did not assume any dramatic increase in longevity, or the creation of any entitlement program to pay for longevity care. If both occur, as gerontological optimists predict, the number of old will skyrocket, as will the cost of maintaining them. To adjust to these "very bad fiscal effects," says the OECD economist Pablo Antolin, one of the report's coauthors, societies have only two choices: "raising the retirement age or cutting the benefits." He continues, "This is arithmetic—it can't be avoided." The recent passage of a huge new prescription-drug program by an administration and Congress dominated by the "party of small government" suggests that benefits will not be cut. Raising the age of retirement might be more feasible politically, but it would lead to a host of new problems—see today's Japan.

In the classic job pattern, salaries rise steadily with seniority. Companies underpay younger workers and overpay older workers as a means of rewarding employees who stay at their jobs. But as people have become more likely to shift firms and careers, the pay increases have become powerful disincentives for companies to retain employees in their fifties and sixties. Employers already worried about the affordability of older workers are not likely to welcome calls to raise the retirement age; the last thing they need is to keep middle managers around for another twenty or thirty years. "There will presumably be an elite group of super-rich who would be immune to all these pressures," Ronald Lee, an economic demographer at the University of California at Berkeley, says. "Nobody will kick Bill Gates out of Microsoft as long as he owns it. But there will be a lot of pressure on the average old person to get out."

In Lee's view, the financial downsizing need not be inhumane. One model is the university, which shifted older professors to emeritus status, reducing their workload in exchange for reduced pay. Or, rather, the university *could* be a model: age-discrimination litigation and professors' unwillingness to give up their perks, Lee says, have largely torpedoed the system. "It's hard to reduce someone's salary when they are older," he says. "For the person, it's viewed as a kind of disgrace. As a culture we need to get rid of that idea."

THE PENTAGON HAS RELEASED FEW STATISTICS about the hundreds or thousands of insurgents captured in Afghanistan and Iraq, but one can be almost certain that they are disproportionately young. Young people have ever been in the forefront of political movements of all stripes. University students protested Vietnam, took over the U.S. embassy in Tehran, filled Tiananmen Square, served as the political vanguard for the Taliban. "When we are forty," the young writer Filippo Marinetti promised in the 1909 *Futurist Manifesto,* "other younger and stronger men will probably throw us in the wastebasket like useless manuscripts—we want it to happen!"

The same holds true in business and science. Steve Jobs and Stephen Wozniak founded Apple in their twenties; Albert Einstein dreamed up special relativity at about the same age. For better and worse, young people in developed nations will have less chance to shake things up in tomorrow's world. Poorer countries, where the old have less access to longevity treatments, will provide more opportunity, political and financial. As a result, according to Fred C. Iklé, an analyst with the Center for Strategic and International Studies, "it is not fanciful to imagine a new cleavage opening up in the world order." On one side would be the " 'bioengineered' nations," societies dominated by the "becalmed temperament" of old people. On the other side would be the legions of youth—"the protagonists," as the political theorist Samuel Huntington has described them, "of protest, instability, reform, and revolution."

Because poorer countries would be less likely to be dominated by a gerontocracy, tomorrow's divide between old and young would mirror the contemporary division between rich northern nations and their poorer southern neighbors. But the consequences might be different—unpredictably so. One assumes, for instance, that the dictators who hold sway in Africa and the Middle East would not hesitate to avail themselves of longevity treatments, even if few others in their societies could afford them. Autocratic figures like Arafat, Franco, Perón, and Stalin often leave the scene only when they die. If the human lifespan lengthens greatly, the dictator in Gabriel García Márquez's *The Autumn of the Patriarch,* who is "an indefinite age somewhere between 107 and 232 years," may no longer be regarded as a product of magical realism.

Bioengineered nations, top-heavy with the old, will need to replenish their labor forces. Here immigration is the economist's traditional solution. In abstract terms, the idea of importing young workers from poor regions of the world seems like a win-win solution: the young get jobs, the old get cheap service. In practice, though, host nations have found that the foreigners in their midst are stubbornly . . . foreign. European nations are wondering whether they really should have let in so many Muslims. In the United States, traditionally hospitable to migrants, bilingual education is under attack and the southern border is increasingly locked down. Japan, preoccupied by *Nihonjinron* (theories of "Japaneseness"), has always viewed immigrants with suspicion if not hostility. Facing potential demographic calamity, the Japanese government has spent millions trying to develop a novel substitute for immigrants: robots smart and deft enough to take care of the aged.

According to Ronald Lee, the Berkeley demographer, rises in life expectancy have in the past stimulated economic growth. Because they arose mainly from reductions in infant and child mortality, these rises produced more healthy young workers, which in turn led to more-productive societies. Believing they would live a long time, those young workers saved more for retirement than their forebears, increasing society's stock of capital—another engine of growth. But these positive effects are offset when increases in longevity come from old people's neglecting to die. Older workers are usually less productive than younger ones, earning less and consuming more. Worse, the soaring expenses of entitlement programs for the old are likely, Lee believes, "to squeeze out government expenditures on the next generation," such as education and childhood public-health programs. "I think there's evidence that something like this is already happening among the industrial countries," he says. The combination will force a slowdown in economic growth: the economic pie won't grow as fast. But there's a bright side, at least potentially. If the fall in birth rates is sufficiently vertiginous, the number of people sharing that relatively smaller pie may shrink fast enough to let everyone have a bigger piece. One effect of the longevity-induced "birth dearth" that Wattenberg fears, in other words, may be higher per capita incomes.

For the past thirty years the United States has financed its budget deficits by persuading foreigners to buy U.S. Treasury bonds. In the nature of things, most of these foreigners have lived in other wealthy nations, especially Japan and China. Unfortunately for the United States, those other countries are marching toward longevity crises of their own. They, too, will have fewer young, productive workers. They, too, will be paying for longevity treatments for the old. They, too, will be facing a grinding economic slowdown. For all these reasons they may be less willing to finance our government. If so, Uncle Sam will have to raise interest rates to attract investors, which will further depress growth—a vicious circle.

Longevity-induced slowdowns could make young nations more attractive as investment targets, especially for the cash-strapped pension-and-insurance plans in aging countries. The youthful and ambitious may well follow the money to where the action is. If Mexicans and Guatemalans have fewer rich old people blocking their paths, the river of migration may begin to flow in the other direction. In a reverse brain drain, the Chinese coast guard might discover half-starved American postgraduates stuffed into the holds of smugglers' ships. Highways out of Tijuana or Nogales might bear road signs telling drivers to watch out for *norteamericano* families running across the blacktop, the children's Hello Kitty backpacks silhouetted against a yellow warning background.

Given that today nobody knows precisely how to engineer major increases in the human lifespan, contemplating these issues may seem premature. Yet so many scientists believe that some of the new research will pay off, and that lifespans will stretch like taffy, that it would be short-sighted not to consider the consequences. And the potential changes are so enormous and hard to grasp that they can't be understood and planned for at the last minute. "By definition," says Aubrey de Grey, the Cambridge geneticist, "you live with longevity for a very long time."

H. ALLEN ORR

Devolution

FROM *THE NEW YORKER*

The trial in Dover, Pennsylvania, in 2005, over the teaching of "intelligent design" theory in the local high school, received worldwide attention. But what is "intelligent design"? H. Allen Orr explains that whatever it is, it is certainly not science.

If you are in ninth grade and live in Dover, Pennsylvania, you are learning things in your biology class that differ considerably from what your peers just a few miles away are learning. In particular, you are learning that Darwin's theory of evolution provides just one possible explanation of life, and that another is provided by something called intelligent design. You are being taught this not because of a recent breakthrough in some scientist's laboratory but because the Dover Area School District's board mandates it. In October 2004 the board decreed that "students will be made aware of gaps/problems in Darwin's theory and of other theories of evolution including, but not limited to, intelligent design."

While the events in Dover have received a good deal of attention as a sign of the political times, there has been surprisingly little discussion of the science that's said to underlie the theory of intelligent design, often called ID. Many scientists avoid discussing ID for strategic reasons. If a scientific claim can be loosely defined as one that scientists take seriously enough to debate, then engaging the intelligent-design movement on scientific grounds, they worry, cedes what it most desires: recognition that its claims are legitimate scientific ones.

Meanwhile, proposals hostile to evolution are being considered in more than twenty states; earlier this month, a bill was introduced into the New York State Assembly calling for instruction in intelligent design for all public school students. The Kansas State Board of Education is weighing new standards, drafted by supporters of intelligent design, that would encourage schoolteachers to challenge Darwinism. Senator Rick Santorum, a Pennsylvania Republican, has argued that "intelligent design is a legitimate scientific theory that should be taught in science classes." An ID-friendly amendment that he sponsored to the No Child Left Behind Act—requiring public schools to help students understand why evolution "generates so much continuing controversy"—was overwhelmingly approved in the Senate. (The amendment was not included in the version of the bill that was signed into law, but similar language did appear in a conference report that accompanied it.) In the past few years, college students across the country have formed Intelligent Design and Evolution Awareness chapters. Clearly, a policy of limited scientific engagement has failed. So just what is this movement?

First of all, intelligent design is not what people often assume it is. For one thing, ID is not Biblical literalism. Unlike earlier generations of creationists—the so-called Young Earthers and scientific creationists—proponents of intelligent design do not believe that the universe was created in six days, that Earth is ten thousand years old, or that the fossil record was deposited during Noah's flood. (Indeed, they shun the label "creationism" altogether.) Nor does ID flatly reject evolution: adherents freely admit that some evolutionary change occurred during the history of life on Earth. Although the movement is loosely allied with, and heavily funded by, various conservative

Christian groups—and although ID plainly maintains that life was created—it is generally silent about the identity of the creator.

The movement's main positive claim is that there are things in the world, most notably life, that cannot be accounted for by known natural causes and show features that, in any other context, we would attribute to intelligence. Living organisms are too complex to be explained by any natural—or, more precisely, by any mindless—process. Instead, the design inherent in organisms can be accounted for only by invoking a designer, and one who is very, very smart.

All of which puts ID squarely at odds with Darwin. Darwin's theory of evolution was meant to show how the fantastically complex features of organisms—eyes, beaks, brains—could arise without the intervention of a designing mind. According to Darwinism, evolution largely reflects the combined action of random mutation and natural selection. A random mutation in an organism, like a random change in any finely tuned machine, is almost always bad. That's why you don't, screwdriver in hand, make arbitrary changes to the insides of your television. But, once in a great while, a random mutation in the DNA that makes up an organism's genes slightly improves the function of some organ and thus the survival of the organism. In a species whose eye amounts to nothing more than a primitive patch of light-sensitive cells, a mutation that causes this patch to fold into a cup shape might have a survival advantage. While the old type of organism can tell only if the lights are on, the new type can detect the *direction* of any source of light or shadow. Since shadows sometimes mean predators, that can be valuable information. The new, improved type of organism will, therefore, be more common in the next generation. That's natural selection. Repeated over billions of years, this process of incremental improvement should allow for the gradual emergence of organisms that are exquisitely adapted to their environments and that look for all the world as though they were designed. By 1870, about a decade after *On the Origin of Species* was published, nearly all biologists agreed that life had evolved, and by 1940 or so most agreed that natural selection was a key force driving this evolution.

Advocates of intelligent design point to two developments that in their view undermine Darwinism. The first is the molecular revolution in biology. Beginning in the 1950s, molecular biologists revealed a staggering and unsuspected degree of complexity within the cells that make up all life. This complexity, ID's defenders argue, lies beyond the abilities of Darwinism to explain. Second, they claim that new mathematical findings cast doubt on the power of natural selection. Selection may play a role in evolution, but it cannot accomplish what biologists suppose it can.

These claims have been championed by a tireless group of writers, most of them associated with the Center for Science and Culture at the Discovery Institute, a Seattle-based think tank that sponsors projects in science, religion, and national defense, among other areas. The center's fellows and advisers—including the emeritus law professor Phillip E. Johnson, the philosopher Stephen C. Meyer, and the biologist Jonathan Wells—have published an astonishing number of articles and books that decry the ostensibly sad state of Darwinism and extoll the virtues of the design alternative. But Johnson, Meyer, and Wells, while highly visible, are mainly strategists and popularizers. The scientific leaders of the design movement are two scholars, one a biochemist and the other a mathematician. To assess intelligent design is to assess their arguments.

MICHAEL J. BEHE, a professor of biological sciences at Lehigh University (and a senior fellow at the Discovery Institute), is a biochemist who writes technical papers on the structure of DNA. He is the most prominent of the small circle of scientists working on intelligent design, and his arguments are by far the best known. His book *Darwin's Black Box* (1996) was a surprise bestseller and was named by *National Review* as one of the hundred best nonfiction books of the twentieth century. (A little calibration may be useful here; *The Starr Report* also made the list.)

Not surprisingly, Behe's doubts about Darwinism begin with

biochemistry. Fifty years ago, he says, any biologist could tell stories like the one about the eye's evolution. But such stories, Behe notes, invariably began with cells, whose own evolutionary origins were essentially left unexplained. This was harmless enough as long as cells weren't qualitatively more complex than the larger, more visible aspects of the eye. Yet when biochemists began to dissect the inner workings of the cell, what they found floored them. A cell is packed full of exceedingly complex structures—hundreds of microscopic machines, each performing a specific job. The "Give me a cell and I'll give you an eye" story told by Darwinists, he says, began to seem suspect: starting with a cell was starting 90 percent of the way to the finish line.

Behe's main claim is that cells are complex not just in degree but in kind. Cells contain structures that are "irreducibly complex." This means that if you remove any single part from such a structure, the structure no longer functions. Behe offers a simple, nonbiological example of an irreducibly complex object: the mousetrap. A mousetrap has several parts—platform, spring, catch, hammer, and hold-down bar—and all of them have to be in place for the trap to work. If you remove the spring from a mousetrap, it isn't slightly worse at killing mice; it doesn't kill them at all. So, too, with the bacterial flagellum, Behe argues. This flagellum is a tiny propeller attached to the back of some bacteria. Spinning at more than 20,000 rpms, it motors the bacterium through its aquatic world. The flagellum comprises roughly thirty different proteins, all precisely arranged, and if any one of them is removed, the flagellum stops spinning.

In *Darwin's Black Box,* Behe maintained that irreducible complexity presents Darwinism with "unbridgeable chasms." How, after all, could a gradual process of incremental improvement build something like a flagellum, which needs *all* its parts in order to work? Scientists, he argued, must face up to the fact that "many biochemical systems cannot be built by natural selection working on mutations." In the end, Behe concluded that irreducibly complex cells arise the same way as irreducibly complex mousetraps—someone designs them. As he put it in a recent *Times* Op-Ed piece: "If it looks, walks, and quacks like a duck, then, absent compelling evidence to the contrary, we have warrant to conclude it's a duck. Design should not be overlooked simply because it's so obvious." In *Darwin's Black Box*, Behe speculated that the designer might have assembled the first cell, essentially solving the problem of irreducible complexity, after which evolution might well have proceeded by more or less conventional means. Under Behe's brand of creationism, you might still be an ape that evolved on the African savanna; it's just that your cells harbor micro-machines engineered by an unnamed intelligence some four billion years ago.

But Behe's principal argument soon ran into trouble. As biologists pointed out, there are several different ways that Darwinian evolution can build irreducibly complex systems. In one, elaborate structures may evolve for one reason and then get co-opted for some entirely different, irreducibly complex function. Who says those thirty flagellar proteins weren't present in bacteria long before bacteria sported flagella? They may have been performing other jobs in the cell and only later got drafted into flagellum-building. Indeed, there's now strong evidence that several flagellar proteins once played roles in a type of molecular pump found in the membranes of bacterial cells.

Behe doesn't consider this sort of "indirect" path to irreducible complexity—in which parts perform one function and then switch to another—terribly plausible. And he essentially rules out the alternative possibility of a direct Darwinian path: a path, that is, in which Darwinism builds an irreducibly complex structure while selecting all along for the same biological function. But biologists have shown that direct paths to irreducible complexity are possible, too. Suppose a part gets added to a system merely because the part improves the system's performance; the part is not, at this stage, essential for function. But because subsequent evolution builds on this addition, a part that was at first just advantageous might *become* essential. As this process is repeated through evolutionary time, more and more parts that were once merely beneficial become necessary. This idea was first set forth by H. J. Muller, the Nobel Prize–winning geneticist, in 1939, but it's a familiar process in the development of human technologies. We add new parts like global-positioning systems to cars not because they're necessary but because they're nice. But no one would be surprised if, in fifty years, computers that rely on GPS actually drove our cars. At that point, GPS would no longer be an attractive option; it would be an essential piece of automotive technology. It's important to see that this process is thoroughly Darwinian: each change might well be small and each represents an improvement.

Design theorists have made some concessions to these criticisms. Behe has confessed to "sloppy prose" and said he hadn't meant to imply that irreducibly complex systems "by definition" cannot evolve gradually. "I quite agree that my argument against Darwinism does not add up to a logical proof," he says—though he continues to believe that Darwinian paths to irreducible complexity are exceedingly unlikely. Behe and his followers now emphasize that, while irreducibly complex systems can in principle evolve, biologists can't reconstruct in convincing detail just how any such system did evolve.

What counts as a sufficiently detailed historical narrative, though, is altogether subjective. Biologists actually know a great deal about the evolution of biochemical systems, irreducibly complex or not. It's significant, for instance, that the proteins that typically make up the parts of these systems are often similar to one another. (Blood clotting—another of Behe's examples of irreducible complexity—involves at least twenty proteins, several of which are similar, and all of which are needed to make clots, to localize or remove clots, or to prevent the runaway clotting of all blood.) And biologists understand why these proteins are so similar. Each gene in an organism's genome encodes a particular protein. Occasionally, the stretch of DNA that makes up a particular gene will get accidentally copied, yielding a genome that includes two versions of the gene. Over many generations, one version of the gene will often keep its original function while the other one slowly changes by mutation and natural selection, picking up a new, though usually related, function. This process of "gene duplication" has given rise to entire families of proteins that have similar functions; they often act in the same biochemical pathway or sit in the same cellular structure. There's no doubt that gene duplication plays an extremely important role in the evolution of biological complexity.

It's true that when you confront biologists with a particular complex structure like the flagellum, they sometimes have a hard time saying which part appeared before which other parts. But then it can be hard, with any complex historical process, to reconstruct the exact order in which events occurred, especially when, as in evolution, the addition of new parts encourages the modification of old ones. When you're looking at a bustling urban street, for example, you probably can't tell which shop went into business first. This is partly because many businesses now depend on one another and partly because new shops trigger changes in old ones (the new sushi place draws twenty-somethings who demand wireless Internet at the café next door). But it would be a little rash to conclude that all the shops must have begun business on the same day or that some Unseen Urban Planner had carefully determined just which business went where.

THE OTHER LEADING THEORIST of the new creationism, William A. Dembski, holds a Ph.D. in mathematics, another in philosophy, and a master of divinity in theology. He has been a research professor in the conceptual foundations of science at Baylor University, and was recently appointed to the new Center for Science and Theology at Southern Baptist Theological Seminary. (He is a longtime senior fellow at the Discovery Institute as well.) Dembski publishes at a staggering pace. His books—including *The Design Inference, Intelligent Design, No Free Lunch,* and *The Design Revolution*—are generally well written and packed with provocative ideas.

According to Dembski, a complex object must be the result of intelligence if it was the product neither of chance nor of necessity. The novel *Moby-Dick,* for example, didn't arise by chance (Melville didn't scribble random letters), and it wasn't the necessary consequence of a physical law (unlike, say, the fall of an apple). It was, instead, the result of Melville's intelligence. Dembski argues that there is a reliable way to recognize such products of intelligence in the natural world. We can conclude that an object was intelligently designed, he says, if it shows "specified complexity"—complexity that matches an "independently given pattern." The sequence of letters "JKXVCJUDOPLVM" is certainly complex: if you randomly type thirteen letters, you are very unlikely to arrive at this particular sequence. But it isn't *specified:* it doesn't match any independently given sequence of letters. If, on the other hand, I ask you for the first sentence of *Moby-Dick* and you type the letters "CALLMEISHMAEL," you have produced something that is both complex and specified. The sequence you typed is unlikely to arise by chance alone, and it matches an independent target sequence (the one written by Melville). Dembski argues that specified complexity, when expressed mathematically, provides an unmistakable signature of intelligence. Things like "CALLMEISHMAEL," he points out, just don't arise in the real world without acts of intelligence. If organisms show specified complexity, therefore, we can conclude that they are the handiwork of an intelligent agent.

For Dembski, it's telling that the sophisticated machines we find in organisms match up in astonishingly precise ways with recognizable human technologies. The eye, for example, has a familiar, cameralike design, with recognizable parts—a pinhole opening for light, a lens, and a surface on which to project an image—all arranged just as a human engineer would arrange them. And the flagellum has a motor design, one that features recognizable O-rings, a rotor, and a drive shaft. Specified complexity, he says, is there for all to see.

Dembski's second major claim is that certain mathematical results cast doubt on Darwinism at the most basic conceptual level. In 2002 he focused on so-called No Free Lunch, or NFL, theorems, which were derived in the late nineties by the physicists David H. Wolpert and William G. Macready. These theorems relate to the efficiency of different "search algorithms." Consider a search for high ground on some unfamiliar, hilly terrain. You're on foot and it's a moonless night; you've got two hours to reach the highest place you can. How to proceed? One sensible search algorithm might say, "Walk uphill in the steepest possible direction; if no direction uphill is available, take a couple of steps to the left and try again." This algorithm ensures that you're generally moving upward. Another search algorithm—a so-called blind search algorithm—might say, "Walk in a random direction." This would sometimes take you uphill but sometimes down. Roughly, the NFL theorems prove the surprising fact that, averaged over all possible terrains, no search algorithm is better than any other. In some landscapes, moving uphill gets you to higher ground in the allotted time, while in other landscapes moving randomly does, but on average neither outperforms the other.

Now, Darwinism can be thought of as a search algorithm. Given a problem—adapting to a new disease, for instance—a population uses the Darwinian algorithm of random mutation plus natural selection to search for a solution (in this case, disease resistance). But, according to Dembski, the NFL theorems prove that this Darwinian algorithm is no better than any other when confronting all possible problems. It follows that, over all, Darwinism is no better than blind search, a process of utterly random change unaided by any guiding force like natural selection. Since we don't expect blind change to build elaborate machines showing an exquisite coordination of parts, we have no right to expect Darwinism to do so, either. Attempts to sidestep this problem by, say, carefully constraining the class of challenges faced by organisms inevitably involve sneaking in the very kind of order that we're trying to explain—something Dembski calls the displacement problem. In the end, he argues, the NFL theorems and the displacement problem mean that there's only one plausible source for the design we find in organisms: intelligence. Although Dembski is somewhat noncommittal, he seems to favor a design theory in which an intelligent agent programmed design into early life, or even into the early universe. This design then unfolded through the long course of evolutionary time, as microbes slowly morphed into man.

Dembski's arguments have met with tremendous enthusiasm in

the ID movement. In part, that's because an innumerate public is easily impressed by a bit of mathematics. Also, when Dembski is wielding his equations, he gets to play the part of the hard scientist busily correcting the errors of those soft-headed biologists. (Evolutionary biology actually features an extraordinarily sophisticated body of mathematical theory, a fact not widely known because neither of evolution's great popularizers—Richard Dawkins and the late Stephen Jay Gould—did much math.) Despite all the attention, Dembski's mathematical claims about design and Darwin are almost entirely beside the point.

The most serious problem in Dembski's account involves specified complexity. Organisms aren't trying to match any "independently given pattern": evolution has no goal, and the history of life isn't trying to get anywhere. If building a sophisticated structure like an eye increases the number of children produced, evolution may well build an eye. But if destroying a sophisticated structure like the eye increases the number of children produced, evolution will just as happily destroy the eye. Species of fish and crustaceans that have moved into the total darkness of caves, where eyes are both unnecessary and costly, often have degenerate eyes, or eyes that begin to form only to be covered by skin—crazy contraptions that no intelligent agent would design. Despite all the loose talk about design and machines, organisms aren't striving to realize some engineer's blueprint; they're striving (if they can be said to strive at all) only to have more offspring than the next fellow.

Another problem with Dembski's arguments concerns the NFL theorems. Recent work shows that these theorems don't hold in the case of co-evolution, when two or more species evolve in response to one another. And most evolution is surely co-evolution. Organisms do not spend most of their time adapting to rocks; they are perpetually challenged by, and adapting to, a rapidly changing suite of viruses, parasites, predators, and prey. A theorem that doesn't apply to these situations is a theorem whose relevance to biology is unclear. As it happens, David Wolpert, one of the authors of the NFL theorems, recently denounced Dembski's use of those theorems as "fatally informal and imprecise." Dembski's apparent response has been a tactical retreat. In 2002 Dembski triumphantly proclaimed, "The No Free Lunch theorems dash any hope of generating specified complexity via evolutionary algorithms." Now he says, "I certainly never argued that the NFL theorems provide a direct refutation of Darwinism."

Those of us who have argued with ID in the past are used to such shifts of emphasis. But it's striking that Dembski's views on the history of life contradict Behe's. Dembski believes that Darwinism is incapable of building anything interesting; Behe seems to believe that, given a cell, Darwinism might well have built you and me. Although proponents of ID routinely inflate the significance of minor squabbles among evolutionary biologists (did the peppered moth evolve dark color as a defense against birds or for other reasons?), they seldom acknowledge their own, often major differences of opinion. In the end, it's hard to view intelligent design as a coherent movement in any but a political sense.

It's also hard to view it as a real research program. Though people often picture science as a collection of clever theories, scientists are generally staunch pragmatists: to scientists, a good theory is one that inspires new experiments and provides unexpected insights into familiar phenomena. By this standard, Darwinism is one of the best theories in the history of science: it has produced countless important experiments (let's re-create a natural species in the lab—yes, that's been done) and sudden insight into once puzzling patterns (*that's* why there are no native land mammals on oceanic islands). In the nearly ten years since the publication of Behe's book, by contrast, ID has inspired no nontrivial experiments and has provided no surprising insights into biology. As the years pass, intelligent design looks less and less like the science it claimed to be and more and more like an extended exercise in polemics.

IN 1999 A DOCUMENT FROM the Discovery Institute was posted, anonymously, on the Internet. This Wedge Document, as it came to be called, described not only the institute's long-term goals but its strategies for accomplishing them. The document begins by labeling the idea that human beings are created in the image of God "one of the bedrock principles on which Western civilization was built." It goes on to decry the catastrophic legacy of Darwin, Marx, and Freud—the alleged fathers of a "materialistic conception of reality" that eventually "infected virtually every area of our culture." The mission of the Discovery Institute's scientific wing is then spelled out: "nothing less than the overthrow of materialism and its cultural legacies." It seems fair to conclude that the Discovery Institute has set its sights a bit higher than, say, reconstructing the origins of the bacterial flagellum.

The intelligent-design community is usually far more circumspect in its pronouncements. This is not to say that it eschews discussion of religion; indeed, the intelligent-design literature regularly insists that Darwinism represents a thinly veiled attempt to foist a secular religion—godless materialism—on Western culture. As it happens, the idea that Darwinism is yoked to atheism, though popular, is also wrong. Of the five founding fathers of twentieth-century evolutionary biology—Ronald Fisher, Sewall Wright, J. B. S. Haldane, Ernst Mayr, and Theodosius Dobzhansky—one was a devout Anglican who preached sermons and published articles in church magazines, one a practicing Unitarian, one a dabbler in Eastern mysticism, one an apparent atheist, and one a member of the Russian Orthodox Church and the author of a book on religion and science. Pope John Paul II himself acknowledged, in a 1996 address to the Pontifical Academy of Sciences, that new research "leads to the recognition of the theory of evolution as more than a hypothesis." Whatever larger conclusions one thinks *should* follow from Darwinism, the historical fact is that evolution and religion have often coexisted. As the philosopher Michael Ruse observes, "It is simply not the case that people take up evolution in the morning, and become atheists as an encore in the afternoon."

Biologists aren't alarmed by intelligent design's arrival in Dover and elsewhere because they have all sworn allegiance to atheistic materialism; they're alarmed because intelligent design is junk science. Meanwhile, more than 80 percent of Americans say that God either created human beings in their present form or guided their development. As a succession of intelligent-design proponents appeared before the Kansas State Board of Education earlier this month, it was possible to wonder whether the movement's scientific coherence was beside the point. Intelligent design has come this far by faith.

D. T. Max

The Literary Darwinists

FROM THE *NEW YORK TIMES MAGAZINE*

The domain of Darwin's theory of evolution has widened of late, moving from biology to psychology to the realm of ideas. And now it's invading literature. D. T. Max previews this new form of literary criticism.

Jane Austen first published *Pride and Prejudice* in 1813. She had misgivings about the book, complaining in a letter to her sister that it was "rather too light, and bright, and sparkling." But these qualities may be what make it the most popular of her novels. It tells the story of Elizabeth Bennet, a young woman from a shabby genteel family, who meets Mr. Darcy, an aristocrat. At first, the two dislike each other. Mr. Darcy is arrogant; Elizabeth, clever and cutting. But through a series of encounters that show one to the other in a more appealing light—as well as Mr. Darcy's intervention when an officer named Wickham runs away with Elizabeth's younger sister Lydia (Darcy bribes the cad to marry Lydia)—Elizabeth and Darcy come to love each other, to marry, and, it is strongly suggested at book's end, to live happily ever after.

For the common reader, *Pride and Prejudice* is a romantic comedy.

His or her pleasure comes from the vividness of Austen's characters and how familiar they still seem: it's as if we know Elizabeth and Darcy. On a more literary level, we enjoy Austen's pointed dialogue and admire her expert way with humor. For similar reasons, critics have long called *Pride and Prejudice* a classic—their ultimate (if not well defined) expression of approval.

But for an emerging school of literary criticism known as Literary Darwinism, the novel is significant for different reasons. Just as Charles Darwin studied animals to discover the patterns behind their development, Literary Darwinists read books in search of innate patterns of human behavior: child bearing and rearing, efforts to acquire resources (money, property, influence), and competition and cooperation within families and communities. They say that it's impossible to fully appreciate and understand a literary text unless you keep in mind that humans behave in certain universal ways and do so because those behaviors are hard-wired into us. For them, the most effective and truest works of literature are those that reference or exemplify these basic facts.

From the first words of the first chapter ("It is a truth universally acknowledged, that a single man in possession of a good fortune, must be in want of a wife") to the first words of the last ("Happy for all her maternal feelings was the day on which Mrs. Bennet got rid of her two most deserving daughters"), the novel is stocked with the sort of life's-passage moments that resonate with meaning for Literary Darwinists. (One calls the novel their "fruit fly." The women in the book mostly compete to marry high-status men, consistent with the Darwinian idea that females try to find mates whose status will ensure the success of their offspring. At the same time, the men are typically competing to marry the most attractive women, consistent with the Darwinian idea that males look for youth and beauty in females as signs of reproductive fitness. Darcy and Elizabeth's flips and flops illustrate the effort mammals put into distinguishing between short-term appeal (a pert step, a handsome coxcomb) and long-term appropriateness (stability, commitment, wealth, underlying good health). Meanwhile,

Wickham—the penniless officer who tries to make off first with Darcy's sister and then carries off Lydia—serves as an example of the mating behavior evolutionary biologists call (I'm using a milder euphemism than they do) "the sneaky fornicator theory."

Humans beyond reproductive age also have a part to play in the Literary Darwinist paradigm. Consider Mrs. Bennet, Elizabeth's mother. Jane Austen calls her "invariably silly," and most critics over nearly two centuries have agreed. But for Literary Darwinists, her marriage obsession makes sense, because she also has a stake in what is going on. If one of her daughters has a child, Mrs. Bennet will have further passed on her genetic material, fulfilling the ultimate aim of living things according to some evolutionary theorists: the replication of one's genes. (J. B. S. Haldane, a British biologist, was once asked if he would trade his life for his brother's and replied no, but that he would trade it for two brothers or eight cousins.)

It is useful to know a bit about current literary criticism to understand how different the Darwinist approach to literature is. Current literary theory tends to look at a text as the product of particular social conditions or, less often, as a network of references to other texts. (Jacques Derrida, the father of deconstruction, famously observed that there was "nothing outside the text.") It often focuses on how the writer's and the reader's identities—straight, gay, female, male, black, white, colonizer or colonized—shape a particular narrative or its interpretation. Theorists sometimes regard science as simply another form of language or suspect that when scientists claim to speak for nature, they are disguising their own assertion of power. Literary Darwinism breaks with these tendencies. First, its goal is to study literature through biology—not politics or semiotics. Second, it takes as a given not that literature possesses its own truth or many truths but that it derives its truth from laws of nature.

The Literary Animal, the first scholarly anthology dedicated to Literary Darwinism, is to be published next month. It draws from the various fields that figure in Darwinian evolutionary studies, including contributions from evolutionary psychologists and biologists as well as literature professors. The essays consider the importance of the male-male bond in epics and romances, the battle of the sexes in Shakespeare and the motif in both Japanese and Western literature of men rejecting children whom their wives have conceived in adultery. *The Literary Animal* spans centuries and individual cultures with bravura, if not bravado. "There is no work of literature written anywhere in the world, at any time, by any author, that is outside the scope of Darwinian analysis," Joseph Carroll, a professor of English at the University of Missouri at St. Louis, writes in an essay in *The Literary Animal.* Why bring literature into what is essentially a social science? Jonathan Gotschall, an editor of *The Literary Animal,* offers an answer: "One thing literature offers is data. Fast, inexhaustible, cross-cultural and cheap."

There is a circularity to an argument that uses texts about people to prove that people behave in human ways. (I'm reminded of the Robert Frost line: "Earth's the right place for love: / I don't know where it's likely to go better.") But Literary Darwinism has a second focus too. It also investigates *why* we read and write fiction. At the core of Literary Darwinism is the idea that we inherit many of the predispositions we deem to be cultural through our genes. How we behave has been subjected to the same fitness test as our bodies: if a bit of behavior has no purpose, then evolution—given enough time—may well dispense with it. So why, Literary Darwinists ask, do we make room for this strange exercise of the imagination? What are reading and writing fiction *good* for? In her essay "Reverse-Engineering Narrative," Michelle Scalise Sugiyama tries to simplify the question by picking stories apart, breaking them down into characters, settings, causalities and time frames ("the cognitive widgets and sprockets of storytelling") and asking what purpose each serves: How do they make us more adaptive, more capable of passing on our genes?

For the moment, Literary Darwinism is a club that may grow into a crowd; there are only about thirty or so declared adherents in all of academia. (The wider field of biopoetics—which relates music and the visual arts to Darwin as well—can claim another handful.) But it has captured the imagination of a number of academics who grew up with other literary critical techniques and became dissatisfied. Brian Boyd, for instance, a well-known scholar of Vladimir Nabokov and professor at the University of New Zealand in Auckland, changed his focus in his forties to Literary Darwinism, gripped by what he calls its "one very simple and powerful idea."

It may seem strange that English professors in search of inspiration would turn to evolutionary biology, but you should never underestimate the appeal of the worldview Darwin formulated. It has a way of capturing people's attention. While not everyone enjoys being reminded that humans descend from monkeys (or even worse, from prokaryotic bacteria), many of us like the subtle reassurance that Darwinism offers. Despite its theory that unceasing change is the essence of life, it can be perceived as a reassuring philosophy, one that believes there are answers. And a philosophy that implies "survival of the fittest" pays a great compliment to all of us who are here to read about it. So it is little surprise that evolutionary biology has come to be invoked not merely as a theory about changes in the physical makeups of living beings but also as an explanatory tool that appeals to both academics and to everyone's inner pop psychologist. (Jack Nicholson explaining his bad-boy behavior to an interviewer for the *New York Times* in 2002: "I have a sweet spot for what's attractive to me. It's not just psychological. It's also glandular and has to do with mindlessly continuing the species.")

Literary Darwinism—like many offshoots of Darwinism—tends to find favor with those looking for universal explanations. Like Freudianism and Marxism, it has large-scale ambitions: to explain not just the workings of a particular text or author but of texts and authors over time and across cultures as well. It may also allow English professors to grab back some of the influence—and money—

that the sciences, in the Darwinian fight for university resources, have taken from the humanities for the past century. But for now, to march under the Literary Darwinist banner you had better be independent and unafraid. "The most effective and easiest form of repudiation is to ignore us," Carroll says.

Literary Darwinists give off a cultlike vibe. When they talk about likeminded academics who won't acknowledge their beliefs in public, they sometimes call them "closeted." The fifty-six-year-old Carroll's own conversion to the discipline took place when, as a young, tenured but disgruntled professor of English at the University of Missouri at St. Louis in the early 1990s, he picked up *On the Origin of Species* and *The Descent of Man* and had an "intuitive conviction" that he had found the master keys to literature. Carroll had always liked big ideas; he'd had a "big Hegel phase" when he was twenty-one. "The basic conception crystallized for me in a matter of weeks," he remembers, and the notes he began taking "at high intensity" formed themselves into the founding text in the field, *Evolution and Literary Theory*, published in 1995.

Jonathan Gottschall, a thirty-three-year-old editor of *The Literary Animal,* began his graduate studies in English at the State University of New York at Binghamton in 1994 and was surprised at how little his professors cared about linking literature with "the big, Delphic project of seeking the nature of human nature. They didn't believe in knowledge. In fact they could only render the word in quotes." When he found a copy of the zoologist Desmond Morris's 1967 book, *The Naked Ape,* in a used bookstore, Morris's observations on the overlap between primate and human behavior spoke to him. (Animals often play a role in these conversion narratives: Ellen Dissanayake, the author of *What Is Art For?* and a biopoeticist at the University of Washington, was primed for her conversion in part by watching the behavior of wild animals—her husband at the time was a director at the National Zoo in Washington—and comparing them to her young children.

Soon after reading *The Naked Ape,* Gottschall reread the *Iliad,* one of his favorite books: "As always," he writes in the introduction to *The Literary Animal*, "Homer made my bones flex and ache under the weight of all the terror and beauty of the human condition. But this time around I also experienced the *Iliad* as a drama of naked apes—strutting, preening, fighting, tattooing their chests, and bellowing their power in fierce competition for social dominance, desirable mates, and material resources." He brought his ideas to class. "When I would say things like 'sociobiology' and 'evolutionary biology' in class," Gottschall remembers, "my classmates would hear things like 'eugenics' and 'Hitler.' It was a measure of how toxic the material was."

His interest in Literary Darwinism does not seem to have helped Gottschall's career—*The Literary Animal* was rejected by more than a dozen publishers before Northwestern University Press agreed to take it on. And Gottschall himself remains unemployed (though that is a condition familiar to many English Ph.D.s). Literary Darwinists claim that no acknowledged member of their troupe has ever received tenure in this country. "Most of my closest friends ended up at the Ivies or their equivalents," Joseph Carroll says, while he is at "a branch campus in a state university system."

THE ALPHA MALE of Literary Darwinism is the seventy-six-year-old Harvard biologist Edward O. Wilson. "There's no one we owe so much," Gottschall says. Wilson contributed a foreword to *The Literary Animal* in which he writes that if Literary Darwinism succeeds and "not only human nature but its outermost literary productions can be solidly connected to biological roots, it will be one of the great events of intellectual history. *Science and the humanities united!*" Wilson has been working for thirty years to prepare the way for such a moment. In 1975 he began the expansion of modern evolutionary biology to human behavior in his book *Sociobiology: The New Synthesis.* In the last chapter, he tried to show that evolutionary pressures play a big role not just in animal societies but also in human culture. "Many scientists and others believed it would have been better if I had stopped at chimpanzees," Wilson would remember later, "but the challenge and the excitement I felt were too much to resist."

In *On Human Nature*, published three years later, Wilson revisited the question with new energy. The field that emerged in part out of his work, evolutionary psychology, asserts that many of our mental activities and the behaviors that come from them—language, altruism, promiscuity—can be traced to preferences that were encoded in us in prehistoric times when they helped us to survive. According to evolutionary psychologists, everything from seasonal affective disorder to singing to lifesaving is—or at least might be—hard-wired. Evolutionary psychologists also try to demystify the nature of consciousness itself, positing, for example, that the brain is a collection of separate modules evolved to serve mental operations, more like a Swiss Army knife than a soul. A controversial implication of their theories is that evolution may be responsible for some inequalities among groups. One has only to recall the trouble that Lawrence Summers, Harvard's president, brought on himself earlier this year when he speculated that evolution might have left women less capable than men of outstanding performance in engineering and science to see how the notion continues to roil us.

All the same, today we speak casually of innate preferences, adaptive behavior, and fitness strategies. Consider how evolutionary psychology has displaced Freud. Who, upon discovering that a remote tribe had an incest taboo, would ascribe it to unconscious repression on the part of the sons of their sexual attraction to their mothers? Instead, we would likely cite an evolutionary biology principle that states that we have evolved an innate repulsion to inbreeding because it creates birth defects and birth defects are a barrier to survival.

In a recent telephone conversation, I asked Wilson to assess the state of the revolution he helped touch off. How far had sociologists and psychologists gone in folding evolutionary principles into their work? Wilson laughed and said silkily, "Not far enough, in my opinion." Nonetheless, he looks forward to seeing sociobiology dust the wings of the arts—especially literature—with its magic. "Confusion is what we have now in the realm of literary criticism," Wilson writes in his foreword to *The Literary Animal.* He amplified the point on the phone: "They just go on presenting it, teaching it, explaining it as best they can." He saw in literary criticism, especially the school led by Derrida, a "form of unrooted free association and an attempt to build rules of analysis on just idiosyncratic perceptions of how the world works, how the mind works. I could not see anything that was truly coherent." Predicting my objection, he went on: "We're not talking about reducing, corroding, dehumanizing. We're talking about adding deep history, deep genetic history, to art criticism."

Literary Darwinists use this "deep history" to explain the power of books and poems that might otherwise confuse us, thus hoping to add satisfaction to our reading of them. Take, for instance, *Hamlet.* Through the Literary Darwinist lens, Shakespeare's play becomes the story of a young man's dilemma choosing between his personal self-interest (taking over the kingdom by killing his uncle, his mother's new husband) and his genetic self-interest (if his mother has children with his uncle, he may get new siblings who carry three-eighths of his genes). No wonder the prince of Denmark cannot make up his mind.

Or look at Jonathan Gottschall's study of the *Iliad,* which emphasizes how the fighting over women in the epic is not the substitute for the fight over territory, as commentators usually assume, but the central subject of the poem, occasioned by an ancient sex-ratio imbalance, a fact he unearthed in part from studies of the archaeological records of contemporary grave sites.

One of the central beliefs of evolutionary psychology is that pleasure is adaptive, so it is meaningful that Literary Darwinism is enjoyable to practice. But while its observations on individual books can be fun and memorable, they also feel flimsy. As David Sloan Wilson, an editor of *The Literary Animal* and a professor of biology and anthropology at SUNY-Binghampton, puts it, "Tasty slice, but where's the rest of the pie?"

And Literary Darwinism is not equally good at explaining everything. It is best on big social novels, on people behaving in groups. As the British novelist Ian McEwan notes in his contribution to *The Literary Animal,* "If one reads accounts of . . . troops of bonobo . . . one sees rehearsed all the major themes of the English nineteenth-century novel." But I don't think even by stretching one's imagination primates evoke *The Waste Land* or *Finnegans Wake.* Tone, point of view, reliability of the narrator—these are literary tropes that often elude Literary Darwinists, an interpretive limitation that can be traced to Darwin himself; his son once complained that "it often astonished us what trash he would tolerate in the way of novels. The chief requisites were a pretty girl and a good ending." Darwin was drawn to books that were Darwinian. Similarly, Literary Darwinists are better on Émile Zola and John Steinbeck than, say, Henry James or Gustave Flaubert. I would read their take on Shakespeare's histories before the tragedies and the tragedies before the comedies, and in *The Tempest* I'd be curious about their observations on the Prospero, Miranda, and Fernando triad but not on Caliban or Ariel. I don't care if there are selection pressures on mooncalfs and sprites.

ULTIMATELY, LITERARY DARWINISM may teach us less about individual books than about the point of literature. But what can the purpose of literature be, assuming it is not just a harmless oddity? At first glance, reading is a waste of time, turning us all into versions of Don Quixote, too befuddled by our imaginations to tell windmills from giants. We would be better off spending the time mating or farming. Darwinists have an answer—or more accurately, many possible answers. (Literary Darwinists like multiple answers, convinced the best idea will win out.) One idea is that literature is a defense reaction to the expansion of our mental life that took place as we began to acquire the basics of higher intelligence around forty thousand years ago. At that time, the world suddenly appeared to *Homo sapiens* in all its frightening complexity. But by taking imaginative but orderly voyages within our minds, we gained the confidence to interpret this new vastly denser reality. Another theory is that reading literature is a form of fitness training, an exercise in "what if" thinking. If you could imagine the battle between the Greeks and the Trojans, then if you ever found yourself in a street fight, you would have a better chance of winning. A third theory sees writing as a sex-display trait. Certainly writers often seem to be prenning when they write, with an eye toward attracting a desirable mate. In *The Ghost Writer,* Philip Roth's narrator informs another writer that "no one with seven books in New York City settles for" just one woman. "That's what you get for a couplet."

Yet another theory is that the main function of literature is to integrate us all into one culture; evolutionary psychologists believe shared imaginings or myths produce social cohesion, which in turn confers a survival advantage. And a fifth idea is that literature began as religion or wish fulfillment: we ensure our success in the next hunt by recounting the triumph of the last one. Finally, it may be precisely writing's uselessness that makes it attractive to the opposite sex; it could be that, like the male peacock's exuberant tail, literature's very unnecessariness speaks to the underlying good health of its practitioner. He or she has resources to burn.

Generally, Literary Darwinism positions literature not as a luxury or as an add-on but as connected with our deepest selves. There is a grandeur to this view, and also a good deal of conjecture. That is because evolutionary biology is unusual among the sciences in asking not just "how" things work but also "why"—and not the why of local explanations (Why does water freeze at 32 degrees?) but the why of deeper ones, why something exists (Why did we evolve lungs? Why do we feel love?) There is no lab protocol to solve these sorts of mysteries, which the inductive techniques of science are poorly designed to answer, and so in the end, evolutionary biologists' conclusions can far outrun their research.

Take, for example, the human fear of snakes. According to Edward Wilson, this fear had its beginning in prehistoric times, when many of our ancestors were killed by snakebites. Those who feared snakes survived in greater numbers than those who didn't. This was the period when the human brain was becoming hard-wired, so our fear, rooted now in our genetic makeup, outlived its usefulness. Even after snakes stopped killing us very often, we remembered how we felt when they did. Over time, because they had traumatized us when we were most impressionable, snakes took a central role in our imaginative lives, becoming a center of our religion and art—whence the protection of the kings of ancient Egypt by the cobra goddess Wadjet; Quetzalcoatl, the Aztec serpent god of death and resurrection; and the fascination D. H. Lawrence felt when an uninvited guest slithered "his yellow-brown slackness soft-bellied" down to his water trough.

It is a nice story backed by some evidence. Children have a readiness to fear snakes that needs only an encounter or two to set it off. Their fear remains even after they outgrow ordinary childhood fears. And many primates, our nearest relatives, also have a readiness—an easily evoked potential—to be afraid of snakes. But we need to know a great deal before asserting that our snake obsession is an example of the sort of "gene-culture co-evolution," in Wilson's words, that evolutionary psychology—and Literary Darwinism—depend on. For one thing, if there is a module in the brain that contains the predisposition to fear snakes, it has not yet been found. Nor do we really know how many snake deaths there were in prehistoric times. Nor whether that number was sufficient to create a phobia, which, moreover, for some reason would have had to remain fixed until the present day in the human mind instead of dropping out through further evolutionary selection, as you might expect a useless phobia to do. Today it might be people who love snakes who outreproduce the ophidophobes, since some snakes make good eating and their skins can be sold for money, yet we have no evidence of this pattern. At the same time, we must ask why there are equivalent or greater dangers our ancestors withstood that do not seem to have led to phobias—for instance, fire.

When you try to evaluate the importance of snakes to myths and the arts, you have to make several more assumptions. First, are snakes any more prominent in our imaginations than, say, eagles, which have never preyed on us? And if they are, does it not seem as likely that our fascination with them comes from there being something special (module-activating, if you like) about the snake's motion or its shape—its resemblance to a stick, or *pace* Freud, to the penis? Or about the fact that it kills with poison rather than through lethal wounding, as most wild animals do? Why trace our fear of them only back to their supposed role as a prehistoric killer of our ancestors?

Sometimes evolutionary psychological theory feels like a start toward a science rather than a science itself. Consider, for instance, the larger question of the human imagination's role in evolution. Let's assume the capacity for imagination is inherited. Then most evolutionary psychologists would assume that human imagination was favored by natural selection and that it helps us to survive. But imagination could just as well not be an adaptation to (imagined) survival pressures but an accidental byproduct of such an adaptation. Maybe evolutionary pressures favored a related mental process like, say, curiosity, and because the higher brain, where such mental activities reside, is a sort of huge pool of neurons, it also produced the capacity for imagination. And, as Stephen Kosslyn, a Harvard psychology professor, notes, "Whether any of this was itself the target of natural selection is anybody's guess."

To be fair, evolutionary psychologists deserve credit for asking whether complex human behavior can be transmitted through a genetic-cultural link even if they cannot yet show that it is. Theirs remains an alluring approach. What they need in order to overcome their problems is the equivalent of the early twentieth-century elaboration of the function of genes—or at least more and better hard science to support their conclusions.

A similar focus would help Literary Darwinists. They would benefit from studying writers and readers in the laboratory to see what parts of the brain our taste for literature comes out of and what the implications are. Such experiments could reveal quite remarkable things. For instance, we know that a structure in the brain called the hippocampus has a key role in long-term memory formulation. Scanning readers using functional MRIs—MRIs set to track blood flow to different areas of the brain—we can also see how different works activate their readers' hippocampuses. Those works that light up the hippocampus the most are the ones people wind up remembering best. So functional MRIs of the hippocampus could provide the beginning of a biological basis for the hoary assumption that *Pride and Prejudice* is a classic and maybe even a justification for the rest of the literary canon.

Even more interesting, brain scanning might one day help to explain the act of reading itself. "Reading is a funny kind of brain state," says Norman Holland, a professor who teaches a course on brain science and literature at the University of Florida in Gainesville. "If you're engrossed in a story, you're no longer aware of your body; you're no longer aware of your environment. You feel real emotions toward the characters." What is going on in our heads? Are we in a dream? A heightened reality? A trance?

Edward Wilson told me that he is confident neurobiology can help confirm many of evolutionary psychology's insights about the humanities, commending the work to "any ambitious young neurobiologist, psychologist, or scholar in the humanities." They could be the "Columbus of neurobiology," he said, adding that if "you gave me a million dollars to do it, I would get immediately into brain imaging." In fact, you won't always need a million dollars for the work, as the cost of MRI technology goes down. "Five years from now, every psychology department will have a scanner in the basement," says Steven Pinker, a Harvard cognitive psychologist. With the help of those scanners, Wilson says that science and the study of literature will join in "a mutualistic symbiosis," with science providing literary criticism with the "foundational principles" for analysis it lacks.

David Sloan Wilson, the co-editor of *The Literary Animal* (and the son of the novelist Sloan Wilson), sees the potential of that embrace differently. "Literature," he says, "is the natural history of our species," and its diversity proves us diverse. No one in *Pride and Prejudice* takes exception when, at the book's opening, Elizabeth Bennet's father's cousin comes to propose to her. In Daniel Defoe's *Moll Flanders,* the title character can, at the same time, consider her incest with her brother "the most nauseous thing to me in the world" and say she "had not great concern about it in point of conscience" because she had not known they were related. Humans are complex, and the best books about them are too. So rather than narrowing literature, David Wilson says that Literary Darwinism may broaden evolutionary psychology.

It may, in fact, have already done so. Think about evolutionary psychology. It is seductive and metaphoric, alluring and imagistic. It is fun to riff on. It takes bits of information and from them builds a worldview. It convinces us that we understand why things happen the way they happen. When it succeeds, evolutionary psychology impresses us with the elegance and economy of that vision and, when it fails, gives us a sense of waste and unthriftiness on the author's part. It may be true or it may just have some truth in it, and once you have encountered it, you can never see things quite the same way again: it works a kind of conversion in you. Isn't it, then, already a lot like literature?

KAREN WRIGHT

The Day Everything Died

FROM *DISCOVER*

Scientific debates can be fierce, and few have been fiercer than the battle raging over the cause of the Permian extinction, the largest wide-scale extinction in Earth's history. Karen Wright reports on the young geologist whose proposed solution to this 250-million-year-old puzzle is causing so much controversy.

One of the boldest assertions ever published in the scientific literature started with a single modest observation. In the late 1970s, geologist Walter Alvarez of the University of California at Berkeley and his father Luis, a Nobel Prize–winning physicist, found an unusual chemical signal in an ancient layer of Italian clay. The clay was enriched in iridium, a rare metal that comes mostly from meteorites, interplanetary dust, and other cosmic debris. The iridium spike appeared in sediments sixty-five million years old, at the so-called K-T boundary between the Cretaceous and the Tertiary periods. It coincided with the demise of the dinosaurs.

Contamination from local sources or a glitch in the iridium-counting machine could have explained the finding. But the Alvarezes found an even bigger spike in another Cretaceous-Tertiary deposit in Denmark. Their interpretation, published in 1980, was heretical.

The clay at the K-T boundary was high in iridium, they said, because it was made of the ash and dust from a six-mile-wide asteroid that had crashed into Earth with the energy of one hundred million megatons of TNT. The impact instantly killed every living thing within hundreds of miles. The animals that weren't incinerated or gassed by fumes froze or starved to death soon after, when dust kicked up by the impact blotted out the sun for more than a year, killing plant life around the globe. Dinosaurs were only the most conspicuous casualty of an epic disaster that eradicated half of all the species on Earth.

"Their idea was met by instant ridicule and derision by most geologists and paleontologists," recalls paleontologist Michael Benton of the University of Bristol in a recent book. It took another decade of evidence gathering, including the documentation of an impact crater off the Yucatán Peninsula, for the impact theory to win acceptance, he notes. Now " 'Extraterrestrial Cause for the Cretaceous-Tertiary Extinction' is considered . . . one of the most influential publications in earth sciences in the twentieth century," Benton writes in *When Life Nearly Died: The Greatest Mass Extinction of All Time.*

The book, however, is not about the Cretaceous-Tertiary impact. And the death of the dinosaurs was not the greatest mass extinction of all time. That superlative belongs to a more severe crisis at the P-T boundary, between the Permian and Triassic periods. Fossil records show that about 250 million years ago, 90 percent of the species on Earth were snuffed out in an abrupt event that spanned the globe. The extinction occurred a couple hundred million years before the dinosaurs died out, so its causes, like its sediments, are buried more deeply. No one has even come close to proving what happened.

But in the past five years, one scientist has dared to implicate a familiar culprit: an asteroid or comet comparable in size and speed to the K-T perpetrator. Geologist Luann Becker of the University of California at Santa Barbara has published a series of papers describing rocks from China, Japan, and Antarctica that have subtle and sometimes unorthodox signs of an impact, including extraterrestrial gases trapped in microscopic carbon cages and minerals deformed by shock waves. Last year, her research team delivered the coup de grâce: evidence of an impact crater off the northwest coast of Australia, hidden beneath two miles of sediment on the ocean floor.

Like the Alvarezes' theory, Becker's Permian extinction work has been greeted with hostility. It prompted a vitriolic exchange in the journal *Science* and a showdown at last December's annual meeting of the American Geophysical Union. NASA has launched an investigation to explore Becker's claims, and some of Becker's peers are second-guessing any findings that fit her interpretation—even if the findings are their own. In January, for example, geologist Peter Ward of the University of Washington in Seattle revised his earlier thesis that the extinction had occurred suddenly, documenting new fossil successions that suggest a more prolonged die-off. Geologist Greg Retallack of the University of Oregon in Eugene is retracting evidence of impact-shocked minerals at the P-T boundary he reported in the late 1990s.

Compared with those researchers, Becker is young and relatively inexperienced, but she cannot be dismissed as a fringe figure. Her academic credentials are impeccable, and she publishes in the country's most prestigious science journal with experts from top-flight universities as coauthors. Although highly qualified scientists often disagree, some insiders are baffled by the heat of this particular debate. Retallack, for one, still believes that an impact scenario is credible. "I don't know why people are trashing Luann," he says.

There could be two reasons: She's wrong, or she's right. If she's wrong, say her detractors, her crusade is drawing focus away from investigators looking at other, more likely scenarios, such as the eruption of hundreds of volcanoes in prehistoric Siberia. "All this putative impact stuff is muddying the waters," grumbles geophysicist Jay Melosh of the University of Arizona at Tucson.

If she's right, then a newcomer who wound up studying the Permian extinction and its "putative impact" has bested paleontological stalwarts who have devoted decades to solving the puzzle of mass extinctions. "I've been all over the world looking for shocked minerals at the P-T boundary, and I haven't found any at all," says geologist Michael Rampino of New York University.

It's the objection of a seasoned scientist, but it could just as easily be the complaint of a runner-up. Becker could be driving a discovery as profound as any in earth science, or she could be courting career-dashing disgrace. Depending on whom you ask, the cause of the greatest extinction of all time has been either finally identified or hopelessly obscured.

TO UNDERSTAND THE FUSS over Becker's claims, it helps to know that ideas about the Permian extinction have long been subject to scholarly caprice. Two centuries ago, the very concept of extinction itself was considered scandalous. The great thinkers of the early 1800s only grudgingly acknowledged that the fossils of mastodons, mammoths, and giant ground sloths unearthed in the previous century had no living counterpart remaining on Earth. Then they portrayed extinction as a gradual event. The preeminent British geologist Charles Lyell maintained that iguanodons, ichthyosaurs, and pterodactyls might stage a comeback if hospitable habitats and climates returned. Lyell also came out against any notion of sudden, indiscriminate cataclysm in the history of life. He branded such catastrophism muzzy-headed voodoo science.

The geologists who defined the fossil hallmarks of the Permian in the 1840s must have feared Lyell's opprobrium, for they failed to mention the signs of mass extinction at the end of that period. It seems unlikely that they simply overlooked it. The Permian extinction obliterated ecosystems as complex as any on Earth today. On land, ten-foot-long saber-toothed reptiles succumbed, and grazing, root-grubbing, and insect-eating lizards vanished, along with the plants and bugs they ate. In the ocean, reefs teeming with life were reduced to bare skeletons. The Permian even finished off the lowly trilobite—perhaps the one celebrity species of the predinosaur era.

Even when geologists finally acknowledged these disappearances in the fossil record, they decided that, while thorough, the Permian die-off had been prolonged. The best estimates had it taking about ten million years, which doesn't seem terribly cataclysmic. A lot can slowly go wrong in ten million years. The climate can grow too hot or too cold; sea levels can rise or fall; the amount of oxygen in the ocean or the atmosphere can change. Most plants and animals are exquisitely sensitive to such shifts, and many might not be able to adapt. But they would die out one by one over millennia, at such a stately pace that a hypothetical human would hardly notice.

Thus Lyell's gradualism continued to prevail, and catastrophic change stayed taboo for most of the twentieth century. That's one reason the Alvarezes' K-T impact theory seemed so radical, even in 1980. It invoked the sort of deus ex machina that Lyell had disparaged, and it conjured improbable images of instantaneous apocalypse.

But for once, an idea that plays well in the tabloids also turned out to be true. Emboldened by that example, geologists began to revisit other scenes of carnage in rock beds around the globe. In addition to the Cretaceous and Permian die-offs, they had identified three other episodes of mass extinction in the past 500 million years (a die-off is considered a mass extinction when 50 percent or more of all species are extirpated from the fossil record). Some specialists could not help but hope that a single uncomplicated cause might explain all five of Earth's great extinctions. "A few years ago, we thought maybe they're all impacts," says Rampino. The pendulum Charles Lyell had pulled far to one side swung back just as far to the other. For a few years, it stayed there. Catastrophe became all the rage.

With the paradigm shifting, geologists admitted they could not prove that the Permian extinction had been gradual after all. Evidence to the contrary began to surface. In the early 1990s, geologists examined a rock section in China that bore the critical fossils of the P-T boundary interleaved with ashy volcanic layers suitable for isotopic dating. Called the Meishan section, this felicitous stratigraphy—along with advances in radiometric methods—allowed researchers to time the extinction better than ever before. In 1998 a Chinese and American group headed by geochronologist Sam Bowring of MIT nailed the date of the Permian extinction at 251 million years ago. A carbon signature in the Meishan section suggested the catastrophe had lasted at most 165,000 years. In other words, it had happened two orders of magnitude faster than the ten-million-year textbook estimate.

Armed with the new time line, fossil experts began weighing in. In a 2000 survey of 333 marine species in the Meishan section, paleobiologist Doug Erwin of the National Museum of Natural History in Washington showed that the extinction happened abruptly in the oceans. That same year Peter Ward documented a sudden die-off of vegetation on land in present-day South Africa. Lines of evidence were converging, and figures kept ratcheting downward. The duration of the Permian extinction went from hundreds of thousands of years to tens of thousands and, finally, to just thousands. Although they couldn't resolve time in terms of days, weeks, or months, many experts came to believe that the whole doomed Permian assemblage—flora, fauna, and foraminifera—might have bought it overnight. The abrupt demise made an impact scenario seem even more plausible. Then Luann Becker came along.

IN 1991 BECKER WAS WORKING toward her Ph.D. at the Scripps Institution of Oceanography in La Jolla, California, when her adviser, Jeffrey Bada, showed her an article about the discovery of a new form of carbon molecule called a fullerene. Fullerenes are hollow, closed lattices shaped like nanoscale soccer balls or geodesic domes (they're also known as buckyballs, after Buckminster Fuller, the dome's inventor). They had first been synthesized in the laboratory in 1985, but some scientists thought they might also be made in space, in the furnaces of stars.

If fullerenes are star dust, Bada reasoned, they could be among the cosmic debris that has fallen to Earth more or less constantly since the birth of the planet. The biggest payloads, of course, would arrive via meteorites. But would they survive an impact? Becker—who was planning to be an environmental geologist—got swept up in Bada's enthusiasm. The two decided to search for fullerenes near known impact craters. They soon found them, in 1993, at an impact site in Canada nearly two billion years old. The molecular cages at the so-called Sudbury site might have been forged on Earth from the intense heat and pressure of the impact or in a common forest fire. Yet in their hollow centers the fullerenes held captive helium gas with an unearthly composition that was distinctive of some meteorites and interplanetary dust.

"We were absolutely taken aback," Becker says. "What was in these little buckyballs was an extraterrestrial signature."

Becker next succeeded in isolating fullerene molecules directly from meteorites. Encouraged that she had found a new way to trace impact events, she joined with geochemist Robert Poreda of the University of Rochester in New York, who had helped develop the technique to find trapped fullerene gases, to look for buckyballs at the sites of mass extinctions. First they found some at the K-T boundary. Then they found some at the P-T, in rocks from the Meishan section and at another site called Sasayama in Japan. In the first of several controversial papers, Becker and her colleagues reported that the P-T fullerenes contained trapped helium and argon gases with extraterrestrial compositions. The helium content of the Sasayama fullerenes, for example, is more than fifty times higher than background levels.

"Thus, it would appear that [extraterrestrial] fullerenes were delivered to Earth at the P-T [boundary], possibly related to a cometary or asteroidal impact event," Becker and her colleagues concluded. "Our results are consistent with recent paleontological studies that now point to a very rapid extinction event."

Becker's fullerene report received guarded praise. True, the notion of alien gases trapped in microscopic carbon cages for millions and even billions of years strains credulity, especially when you imagine the force of the impact that supposedly delivered them. But by the time Becker's work appeared, impact geologists were sorely in need of alternative tracers. Their two favorites from the K-T days—iridium spikes and shocked quartz—hadn't turned up in any incriminating abundance in the rocks associated with other mass extinctions. So, fullerenes from outer space? Why not? "They looked like a possible winner in terms of a signature of an impact," says Rampino, a coauthor of that first report.

Two years later, Becker and geochemist Asish Basu of the University of Rochester published another paper with still more unconventional evidence for a Permian impact. Becker's group claimed to have found dozens of actual fragments of meteorites in rocks from the P-T boundary in Antarctica. That evidence is unconventional because meteoritic remains are so easily turned to dust. If they had somehow avoided being incinerated on entry or pulverized on impact, they would have disintegrated in a year of heavy rain—long before geologic processes could fold them into native rock. Less than half a dozen meteorite fragments have been found intact in rock layers the world over.

Becker's fragments are intact and unweathered, although they are supposed to be a quarter of a billion years old. "The meteorite fragments . . . are so well preserved that their preservation must be due to rather unusual circumstances," the authors themselves concede. But as far as they are concerned, "the two largest mass extinctions in Earth history at the K-T and P-T boundaries were both caused by catastrophic collisions with chondritic meteoroids."

Seven months later, *Science* published Becker's report of the proposed impact site. This time Becker, Basu, and four other coauthors described a submarine hump called Bedout High that is buried in ocean sediments one hundred miles off the northwest coast of Australia. Geologist John Gorter of ENI Australia was surveying for offshore oil there when he spotted the plateau on a seismic profile of the seabed in the late 1990s. Becker hadn't learned of Gorter's find until 2002, but when she called him he said he could also get her rock cores drilled from the top and the flank of the structure's uplifted bull's-eye. "I got my rear end over there and started looking at those samples," she says.

In those seafloor samples, Becker's team reports finding shocked and melted minerals and glass that could be produced only by the intense heat and pressure of a bolide, or meteoric, crash. The researchers dated one of the mineral grains and got a familiar number: 250 million, give or take a few million. They say a gravity model of the site, a kind of topographical map of buried geologic structures, looks much like the gravity model of Chicxulub, the K-T impact structure. Becker and company say the signs of impact at Bedout are compelling enough to warrant further scrutiny. And they are getting it.

MAYBE IT'S BECAUSE none of her coauthors are Nobel Prize winners. Maybe it's because she and her colleagues are the only ones who know how to find a fullerene. Maybe it's because evidence of an extinction four times older than the K-T is that much more difficult to find and interpret. For whatever reason, Becker's latest paper—"as spectacular and annoying to some people as the Alvarez paper in the 1980s," she says—has fared no better than that historical example. Except there's no sign of eventual acceptance—even from a former coauthor and fan of the impact theory.

"The dates are not unequivocally two hundred fifty million years, the shocked minerals don't look like shocked minerals, and the gravity anomaly doesn't look like the gravity anomaly you'd get from an impact," Rampino observes. "There's no evidence of a crater, let alone a crater of that time period."

Becker's critics have aired their grievances in caustic missives to *Science.* A group led by British sedimentologist Paul Wignall of the University of Leeds writes about examining rocks cored from basin sediments six hundred miles south of Bedout. "At no level in the core . . . is there evidence for a layer of impact ejecta or a tsunamite," the authors contend. Becker's team responds that Wignall's core hasn't even been proved to include material from the Permian-Triassic boundary. Another group headed by geochronologist Paul Renne of the Berkeley Geochronology Center in California notes that the gravity map of Bedout bears no resemblance whatsoever to another confirmed impact site called Vredefort. Becker admits that her gravity signature is a little irregular: In the map's caption she calls it "significantly reduced and more subdued" than Chicxulub. But she says the reference to Vredefort—which is plainly visible on the surface of the South African desert—just shows how irrational her critics have become: "Comparing a crater that's exposed at the surface to something that's been buried under four kilometers of debris? Give me a break. I mean, hello!"

Renne and other investigators also charge that Becker's supposedly shocked minerals don't have the telltale patterns of impact-induced features: narrow, parallel bands crisscrossing at various angles, like a microscopic tartan weave. Rampino says Becker's group offers nothing nearly as persuasive as the shocked minerals found in K-T boundaries across the globe in the years following the Alvarezes' breakthrough paper. He still remembers the day in 1983 when he saw the first slides of K-T shocked quartz at a meeting. "I went down the hall to see it, and I came back convinced," he recalls. "Had there been a picture like that in [Becker's] paper, these questions wouldn't have come up at all."

Additional questions surround Becker's impact glass, which Renne and others believe could be volcanic. In a recent online analysis, earth scientist Andrew Glikson of Australian National University asserts that Bedout is probably just a buried, burned-out volcano. The oil prospectors who originally collected the Bedout cores also assumed that the rocks were volcanic. But everyone thought Chicxulub was a volcanic crater, too, Becker says, until tests done in the late 1980s proved otherwise.

"It's tectonically and geologically impossible for it to be [a volcanic crater]," she claims. "At the time this thing formed, it was in the middle of a basin that was nowhere near a subduction zone—it was nowhere near the kind of geologic activity that would cause a volcano to form."

Even the fullerene tracers that were once warmly received have come under attack, because geochemist Ken Farley of Caltech in Pasadena found no helium in P-T rocks from Meishan when he tried to replicate Becker's work.

"Nobody can reproduce her results," says Melosh, who maintains that Becker's means of isolating fullerenes could also be used to synthesize them. "Possibly she's fooling herself because she's making the fullerenes she's detecting."

To each of these charges, Becker has detailed and spirited retorts. She points out that Farley, for example, did not examine the same Meishan rock samples she did and that he looked for helium in bulk rather than isolating fullerenes first and then looking for gases trapped within them. "We've got everybody hounding us because it's a spectacular claim," says Becker. "They feel threatened. Why else would they make such absurd statements?"

AT THIS POINT, it is reasonable to wonder how an extraterrestrial bolide the size of Mount Everest could plow into the planet without leaving an unambiguous trace. The answer is, in part, that 250 million years of heat and pressure can deal its own damage to rocks. The ocean floor, for example, recycles on a tectonic conveyor belt every 200 million years or so, erasing all signs of disturbance. Bedout sits just offshore on a continental shelf; otherwise its features—whether volcanic or extraterrestrial—would be history. Still, even that relatively stable continental crust erodes, subsides, uplifts, and deforms over the millennia, obscuring its original mien.

So scientists are left to reconstruct epochal events from nearly inscrutable remains. Retallack and others, for example, have found an iridium blip at the P-T boundary, but it's one-tenth as large as the iridium spike reported by the Alvarezes and others at the K-T. That could imply a modest-size meteorite, not big enough to cause a worldwide extinction. But some meteorites contain very little iridium, and comets, which are mostly ice, don't have any. If an impactor landed in the deep ocean, it wouldn't create much shocked quartz either, because the ocean floor has less quartz in it than continental crust. If a king-size comet landed in the deep ocean, it would be like stabbing a man with an icicle: a murder with a weapon that vanishes.

There are other suspects. One is the Siberian Traps, a million-year-long volcanic eruption that flooded five time zones of Russia with basalt lava more than a mile deep. Over the past decade, ever more sophisticated dating of the ancient basalt has shown that the lava bed could be about the same age as the extinction, and recent studies have revealed that it covered twice as much area as previously supposed. "Knowing that this province was probably twice as big as we thought has a visceral effect," says Renne. "We're staring one of the significant coincidences [of the P-T boundary] right in the face."

A million years of eruption might release massive clouds of sulfurous gases and carbon dioxide. "It probably wouldn't have been a lot of fun to breathe," says Renne. The oxidation of coal beds beneath the magma could release methane as well. The sulfur could produce torrents of acid rain, the carbon dioxide and methane could lead to rapid greenhouse warming, and life on Earth just might not have been worth living for a while.

The Siberian Traps hypothesis has been a favorite among Permian experts when times get tough with the impact theory. There has even been speculation that an impact caused the eruptions. But no one has described a convincing mechanism for an impact-induced eruption, especially not one lasting a million years. And some geologists now question whether the eruptions could have been disastrous enough to account for a global extinction. Basaltic eruptions are mild, like those in Hawaii, not spectacular like the pyrotechnic Mount Saint Helens.

"There are no big explosions," says Melosh. "It's very bad if you happen to be right under the lava." As for exterminating life elsewhere on the planet, "I don't think it's enough."

The one element of the Permian mystery that is certain is that something did indeed claim the lives of nine out of every ten species. The fossil signature of the Permian remains the only obvious signal of what went down 250 million years ago, but it, too, resists deciphering. Ward's latest findings are a case in point: Though his 2000 report on South African plant fossils showed signs of an abrupt extermination at the P-T boundary, his new analysis of animal fossils suggests that a gradual extinction preceded that ultimate burst of fatalities.

"Different organisms have different reactions to different stresses, so if you knew the sequence of mortality you could get a handle on the sequence of events," says Renne. Unfortunately, current dating methods aren't precise enough to determine the exact order in which species disappear from the fossil record.

It does seem certain that the extinction was followed by an extraordinarily long recuperation, called a survival interval, of at least four million years. During that time, the fossil record shows that a handful of plants and creatures held on for dear life: humble things such as clams, "well adapted to living in lousy environments," says Erwin. According to Benton, it would be fully one hundred million years before the planet recovered the same level of biodiversity it had hosted before the end-Permian crises.

What then follows is oddly reminiscent of Lyell's dubious notions on the resurrection of extinct forms. Paleontologists have documented a number of plants and animals that disappear at the end of the Permian, stay gone for millions of years, and reemerge in the middle Triassic. They call these Lazarus species.

It's difficult to build a convincing theory of mass extinction around such data. "I think there was definitely an impact," says Retallack, "but I don't think it caused the extinction necessarily." Instead, Retallack imagines that the impact released methane stored in the seafloor when it struck. The methane essentially suffocated life. "The actual mode of death would've been coughing up a blood-specked frothy sputum," says Retallack.

Rampino detects a note of desperation in such scenarios. "The search is widening from the standard cast of characters," he says. "They're pulling suspects off the street. It would be so much easier if we could just find a big crater in the ground, or a big, smoking volcano."

But Renne, for one, would not mourn the loss of the impact theory or any other pat explanation for biocatastrophe. "Why should every major extinction have the same cause?" he says. "It would just be too tidy."

There is nothing tidy about the P-T impact theory as it stands today. "Everybody is waiting for an ending to the story," says Becker.

Well, not everybody. Some mass-extinction veterans, weary from fifteen years of fruitless rock-wrangling at the Permian boundary, are throwing in the trowel.

"I really enjoyed the P-T field in the 1990s," says Retallack. "Then it was fun. Now it's gotten to be name-calling and acrimony. I don't relish the kind of debates that will go on from here on out. I don't intend to pursue further studies of impact tracers." He says he would rather study Paleozoic paleosols—really, really old dirt.

Doug Erwin has a book in press that will bid his adieu to the Permian problem. When he wrote his first book on the extinction in 1993, few of his colleagues were interested. Then "everybody who was busy worrying about the K-T extinction got bored and decided to come down to the P-T," he says. Now it's just too crowded to be any fun. Rampino agrees: "You have to take a number to get to study any boundary. It's like going to the grocery store."

Those players still in the game can look forward to more fireworks later this year, when results from a NASA-funded effort to verify Becker's findings are due to be released. Last fall, the NASA program sent Erwin, Becker, and Frank Kyte, a geologist at UCLA, to the Meishan section in China, where they could decide, on site and in person, which rocks to analyze and how to divvy them up. The rocks were distributed among several U.S. laboratories for independent testing.

"The expedition to do some definitive sampling is just what we need," says Melosh.

And so the crisis in end-Permian science will continue a while longer, claiming careers and ravaging reputations. It remains to be seen whether Luann Becker and her putative impact will be survivors, casualties, or one of those mysteriously resurgent Lazarus species.

JACK HITT

Mighty White of You

FROM *HARPER'S*

Who were the first Americans? Were they people from Asia who crossed a land bridge between Siberia and Alaska twelve thousand years ago and then moved their way down the continent? Recently there have been challenges to this account—with some hypothesizing a pre-Amerindian settlement by Europeans. Jack Hitt finds that these new theories have less to do with science than with a distressing and not-so-subtle racism.

I. Charlemagne's Heir

I WAS SEVENTEEN YEARS OLD when I discovered that I was the great-great (and forty-six more of those) grandson of Charlemagne—king of the Franks and Holy Roman Emperor. Where I grew up, it's not unusual to find out such things. The culture of Charleston, South Carolina, is built around the pride associated with a handful of family histories. Like most of my friends downtown, I spent my youth in an unconscious state of genealogical questing. Might I be the descendant of a signer of the Declaration? Robert E. Lee's messenger? I bugged my mom and aunts and uncles. Who am I really? Might my childhood friends turn out to be third cousins? In Charleston, that one's almost too easy.

My mother grew exhausted with my pestering and sent me to see Mary Pringle, a cousin who was said to spend her days studying the family genealogy. Primed with curiosity, I arrived at Cousin Mary's elegant antebellum home on a hot summer day. After some iced tea and pleasantries, I was presented with a large, unwieldy sheet of paper bearing a set of concentric circles. In the center, Mary wrote my name, and in the next circle, divided in half, she wrote the names of my parents. In the next circle, divided now into fourths, she wrote the names of my four grandparents. We filled it out as far as we could in every direction, and in that area where her family and mine converged—her life's work—a seemingly unbounded wedge flew backward to Scotland and England, until my ancestors were hobnobbing with William Shakespeare and Mary, Queen of Scots. *This line,* she said, pointing to one of the ancient British earls we could claim, *leads in a direct line all the way to Charlemagne.*

This was almost too much past to absorb and too much pride to possess. I wanted to ask her what the Holy Roman Emperor had left me in the will. But Mary's tone was solemn, nearly religious:

You are the direct descendant of King Charlemagne.

The room felt still as the rest of the universe slowly turned on its gyre about me, just as it did on the paper.

I left Mary Pringle's house feeling pretty, well, rooted. It's an important feeling for most people—knowing where they come from. And being heir to Charlemagne would serve me just fine on the gentlemen's party circuit. Over the next few years, I became as cunning at hefting this lumbering chunk of self-esteem into passing conversation as a Harvard grad is at alluding to his alma mater.

Roots are crucial to us—us being all Americans—because they are the source of so much of our national anxiety about not quite belonging. Has any passenger manifest been more fretted over than the *Mayflower*'s? One of the few Internet uses that seems able to compete with porn is genealogy. The most significant television miniseries—*Roots*—spawned a wave of pride among African Americans (and, arguably, even that compound name) and is partly responsible for the ongoing effort to drain the word "white" of its racist intimations by redefining it as "Irish American," "Scottish American," "Italian American," and the like. For everyone—including Native Americans, who itchily remind the rest of us that they might also be called First Americans—there is a deep anxiety about rootedness and its claims. After Bill Frist was elevated to majority leader of the Senate, he self-published a book. Its title cries out as much with this anxiety as it does with pride: *Good People Beget Good People: A Genealogy of the Frist Family.*

And the truth is, this anxiety can never quite be quelled. About three years after I had tea with Mary Pringle, I was in a college calculus class when the teacher made a point about factoring large numbers. He dramatized it by giving an example from the real world, explaining how redundancy affected genealogy. He noted that if you run your line back to AD 800, the number of direct ancestors you would have is preposterously large (today, it would be 281,474,976,710,656, or a quarter quadrillion). Since the total human population for all time is estimated at a sparse 106 billion, the huge number makes no sense unless there is massive redundancy far back in time.

The upshot, the teacher explained, is that nearly everyone currently living anywhere on the planet can claim (and he paused for emphasis) . . . *to be the direct descendant of Charlemagne.*

The room felt still, as the rest of the universe slowly turned on its gyre about me, laughing.

Not long afterward, I learned from Alex Shoumatoff's book *The Mountain of Names* that this paradox has a name, "pedigree collapse," which explains how the old practice of cousins marrying creates super-redundancy in the deep past (and why the planet's total population of 268,435,456 in AD 1300 roughly equals the number of ancestors you would have had at that time). A while back, my aunt Mary died, but her dream lives as one of Amazon's favorite genealogy books: *The Everything Online Genealogy Book: Use the Web to Discover Long-Lost Relations, Trace Your Family Tree Back to Royalty, and Share Your History with Far-Flung Cousins.*

Recognizing the delusional fiction of it all, I swore off distant genealogy forever. That is, until recently, when I learned that new technologies and laboratory breakthroughs have revealed that my great-great (and 638 more of them) grandfather was the first man to set foot on the continent of North America, some sixteen thousand years ago.

II. The Allegory of the Cave

ON A COOL LEAFY HILLSIDE above a trickling Cross Creek in remote Pennsylvania, the sun creeps through the trees, primordial. Nestled into the slope above, an open rock-shelter seems just the place where any self-respecting *Homo sapiens* might set down his basketry and spear and light a fire.

Today there's a parking lot at the hill's base and a set of sturdy stairs that lead to a wooden enclosure built by James Adovasio. He's the Mercyhurst College archaeologist who's been excavating this controversial site since the mid-1970s. Adovasio is guiding a rare tour for a dozen or so amateurs, myself included. He arrives in full archaeological drag: sleeveless flak jacket, boots, work pants, mystical belt buckle. His broad scowling face gives him the look of Martin Scorsese and George Lucas's love child.

Inside the shelter, there's an office, electricity, good lighting, and a suspended boardwalk so that visitors and workers don't stomp over all the evidence. Enormous squared-out holes plunge down into the dense earth, where tiny round markers dangle like pinned earrings in the stone.

It was here that Adovasio found his controversial evidence, stone tools that carbon-date to 16,000 ± 150 years BC.

This makes them at least four millennia older than the last ice age, after which the first humans were traditionally believed to have arrived in North America. Adovasio asks us to notice a pencil-thin black line in the stone. No one can really see it. So Adovasio splashes water on it and the line darkens into little more than a pencil swipe across the rock.

"This is a fire pit," he declares. All of us move closer to the rail to squint and then decide, as with so much of prehistoric archaeology, that we'll just take his word for it. He describes the scene that once occurred here. Folks sat around the fire and cooked deer and squirrel while snacking on hackberries and nuts. Maybe they battered some rocks into spear points or wove some grasses into primitive baskets. In the chilly rock-shelter, it is easy to look around and imagine this ancient gathering. Typically, the prehistoric picture show that plays on the cave wall of our minds involves bandy-legged men with spears pursuing mastodons. Here we are in the kitchen, where people sat around the fire, eating and talking. Away from the picturesque hunt. Quiet time, culture time, story time.

Now, 16,000 ± 150 years later, we are once again gathered here for story time. But Adovasio is not alone in trying to tell this story. Helping him, sort of, is the fat guy in front of me. He's just one of the crowd, like me, but he has spent much of the tour loudly explaining—allegedly to his long-suffering girlfriend but really to the confederacy of dunces that is the rest of us—just how much he knows about this place. He's wearing a fanny pack the size of a car tire cinched above pastel shorts, robin's-egg-blue socks, and black tennis shoes. His XXL T-shirt declares, KLINGON ASSAULT GROUP.

He has already sneeringly uttered the phrase "politically incorrect" several times to signal that he is not a victim of conventional wisdom but a man of daring opinions. He has let everyone in the place know that he very well intends to ask Adovasio the tough questions. And now the time has come: "Professor Adovasio, does working here in the rock-shelter in western Pennsylvania keep you safe from resentments with Native Americans?" He makes an interrogative honking noise.

"No," Adovasio insists. "Native Americans have an intense interest in this site." Adovasio segues quickly into a shaggy-dog story about a certain Indian who was nothing but supportive. I look at the dozen or so of us, all white folks in their forties and fifties, and none of us seems a bit mystified about why Native Americans might be resentful. Perhaps that is why Adovasio doesn't feel it necessary to address this issue. His work, after all, implies that the Native Americans were latecomers, that before Asians crossed the Bering Strait and began settling North America around the commonly agreed time of thirteen thousand years ago, there was already somebody here. He also knows that other scientists claim to have fresh evidence suggesting that these earliest people, and hence the true First Americans, were, in the scientific jargon, "Caucasoid." That is, white people who looked just like the Klingon ± two hundred pounds.

III. American Genesis

Our continent's creation story about the Asian hunter/gatherer crossing the Bering Strait is only about a century old and owes its origin to a black cowboy named George McJunkin. A former slave, McJunkin went out West, taught himself book learning, and herded cattle while pondering the world around him. He was said to ride a horse fixed with a big rifle holster that toted his telescope.

McJunkin was studied-up enough to know that some old bones he found near Clovis, New Mexico, in 1908 belonged to extinct animals. Twenty-five years later, experts investigating McJunkin's discovery found embedded in some of these ancient bison bones a flat, rounded arrowhead with a bit of fluting at the base to assist its fastening to a spear. It would eventually become known as the Clovis point—the oldest spearhead type ever found on the continent.

What makes the Clovis point so special is that it is found in massive numbers all across the continent and reliably enough at a level where organic material generally carbon-dates to roughly twelve thousand years ago. How massive? Take Bell County, Texas. The area north of Austin—known as the Gault site—must have been a well-known pit stop among the Clovis tribes. The place has yielded more than half a million stone artifacts from the Clovis era.

"The whole idea of archaeology is that there must be enough redundancy in the record," said Richard Burger, a professor of anthropology at Yale. Why? Because there is no other way to prove the case in archaeology, no other path to certainty.

"Archaeologists can't do experiments," Burger said. "Unlike lab science, we can't mix carbon and sulfur and conclude that such and such happens. So we have something else approaching that. We take advantage of redundancy, so that the evidence repeats itself in broad patterns. With Clovis, this happens with confidence."

In the last two decades, though, the confident tellers of Clovis Man's tale have been challenged by academic renegades devoted to identifying a new "First American." There are at least four major sites (and some minor ones) in the Americas that claim to have found man-made objects dating to tens of thousands of years before Clovis time. These theorists argue that although Clovis Man might still have crossed the Bering Strait thirteen thousand years ago, there is evidence that somebody else was already here. Given their natural caution, academics generally stop right there.

In the meantime, though, other theorists have stepped forward to identify the pre-Clovis somebody. This process has happened not on the front page of the newspapers but in the rumor mill at the edge of archaeology and anthropology and in the pages of popular-science magazines. The new theory holds that Caucasians from Europe settled this continent first and that Native Americans are just another crowd of later visitors, like Leif Ericson's Vikings and Christopher Columbus's Spaniards. Most important, the way this theory has seeped out of cautious academia and into the pop culture as slightly naughty fact suggests that America's neurosis about race has taken up a new and potentially toxic location—deep in the heart of our continent's creation myth.

IV. Sesquipedalianismo

For any new story to get told, there has to be an opening, a sudden tectonic jarring of a discipline's conventional wisdom. Thomas Kuhn described this critical moment with the now much weathered phrase "paradigm shift." It's the precise moment of the tilt between an old worldview and a new one. And that's where we are now in the subdiscipline of ancient-American archaeology, poised between those views held (as always) by mossbacked conservative traditionalists on the one side and young agitated revolutionaries on the other.

The voice of skepticism and orthodoxy is best embodied by Professor C. Vance Haynes of the University of Arizona. He comes by his skepticism honestly. He once bought into a claim quite similar to Adovasio's, back in the 1960s, at a site called Tule Springs in Nevada. He, too, thought the Clovis line had been breached. He was convinced by extensive evidence of "hearths" filled with charcoal and animal bones, revealing a human encampment dating back twenty-eight thousand years. But later, when Haynes conducted precise tests of the charcoal, he realized that it was merely very old wood turning into coal. All of it was wrong. "You begin to see how easy it is to misinterpret things," Haynes said.

When you look at the evidence and the fights around it, you can understand why. First, arrowheads are cool things. Every little kid who has ever dug one up knows this. To hold in your hand a weapon that is five hundred, a thousand, five thousand years old is humbling and, just, neato. Arrowheads are symmetrical and beautiful objects. Their flutes, their chipped edges, their flared tails have all been studied, categorized, and given handsome names, dozens of them. The Madison point dates from AD 1400, the Whitlock point from 400 BC, further back to the Haywood point at 5000 BC, deeper still to the Cascade point at 8000 BC (or so), and finally to the oldest, the Clovis point.

With each older style, the artifacts become less *arrowheady*. Instead of having tooled edges, they are clearly flintnapped—i.e., beat at with another stone. A beveled edge might be replaced by a straight blade. Those graceful fishtails disappear, and then you get a simple stone point with a groove banged out at the bottom, the telltale primitivity signifying Clovis time. Beyond that, it is hard to tell whether the evidence is, in fact, man-made. Archaeology has a term for naturally occurring objects that appear to be artifacts: geofacts.

The archaeology establishment believes that the entire array of pre-Clovis evidence is a pile of geofacts. And not all that big a pile either. Whereas all the evidence of Clovis Man would crowd a boxcar, all the good physical evidence of pre-Clovis Man could probably fit in your bureau drawer. And when you look at the individual artifacts themselves, the evidence can be pretty underwhelming. There are no broad patterns, there are no similarities, and there is no redundancy.

The dates are a mixed bag, ranging as far back as fifty thousand years ago to more recent sixteen-thousand-year dates. And obtaining those dates is a messy business. Rocks cannot be carbon-dated. Only the organic material they are found nested in can be, and that material is easily contaminated by rain, by burrowing animals, by time. And carbon dating may sound precise, but the idea of it—that carbon 14 molecules throw off electrons at a metronomically consistent geological pace—is more exact than the reality. Almost since the discovery of carbon dating, scientists have been noting phenomena that cause variations with the regularity of carbon's internal clock—sunspots, stray comets, atomic bombs—such that it requires applying a "correction factor." Thus, for any ancient evidence to be confirmed, the punk rockers of archaeology have to look for affirmation from their elders, the Lawrence Welk orchestra. Worse, the old fogies, like Vance Haynes and others, are essentially being asked to confirm a theory that overturns their entire life's work. This combination of murky evidence and professional oedipalism can mean only one thing: academic food fight.

In prehistoric archaeology there's a lot of dialogue between the conservative traditionalists and the rebel theorists that, boiled down, typically goes like this:

Upstart Archaeologist: "This is a primitive stone tool that's sixteen thousand years old."

Eminence Grise: "No, it's not."

Upstart Archaeologist: "Fuck you."

Actually, that's not much of an exaggeration. In Adovasio's book, *The First Americans,* he quotes a friend who said, " 'If they don't believe the evidence, fuck 'em'—definitely not scientific discourse but not ill considered, either!"

From its opening line—" 'Damn,' I said."—Adovasio's book quivers with the fury of a scolded teenager. His own site, the Meadowcroft Rock-shelter I visited in southwest Pennsylvania, has been roundly dismissed by elders who note the existence of nearby "coal seams" (yet another factor that throws off carbon dating) and ground-water seepage. C. Vance Haynes is among those who have wrinkled their noses at Meadowcroft. In his book, Adovasio dismisses Haynes as the "grinch of North American archeology." In fact, anyone who has questioned Adovasio's site at Meadowcroft is generally referred to as a "gnat."

And no love is lost among the rebels themselves. When a Parisian archaeologist discovered an amazing site called Pedra Furada in Brazil, the initial reports were breathtaking. Besides numerous pieces of pre-Clovis evidence, there were cave paintings said to be even older than the images at Lascaux in France or at Altamira in Spain. Some of the Pedra Furada drawings are said to depict hot Pleistocene-era group sex. Brazil's government developed plans to capitalize on the site as a tourist attraction. Adovasio himself was part of the expert panel that, sorrowfully, declared it all wrong. Adovasio wrote that he saw nothing but "almost surely broken rocks that had fallen into the rockshelter." He dismissed the find of "ancient fireplaces" as "nothing more than material blown in from nearby forest fires."

Of course, the language of this brawl is usually academic and Latinate, mostly fought with the manly sesquipedalianisms of science jargon. Here, tree rings are "dendrochronological samples." A rock is a "lithic," and a rock that's clearly been flaked by human hands is "an indubitable lithic artifact." Bits of stone chipped off to make a tool are "percussion flakes."

These are the lyrics of the trade, played in the key of high-science formality. And it's with such swaggering sesquipedalianismo that an entire career of work can be cattily dismissed: "My review has raised doubts about the provenience of virtually every 'compelling,' unambiguous artifact. . . ," wrote the archaeologist Stuart Fiedel in 1999 of the most promising pre-Clovis site ever.

The archaeologist whose work is being trashed here is Tom Dillehay of the University of Kentucky. He had claimed to find—at a Chilean site called Monte Verde—fantastic evidence of a pre-Clovis community: a series of huts, one of which might have been some kind of primitive drugstore, as it contained traces of pharmaceutical herbs. He found a tent post still staked in the ground with knotted twine from a Juncus tree, or, in the jargon, "the indisputably anthropogenic knotted Juncus."

In 1997 a team of specialists, including C. Vance Haynes, visited the site, examined all of Dillehay's cool evidence, and unanimously approved it. The pre-Clovis line was officially breached. Tom Dillehay was the man, though not for long. Haynes began to waver. Then Fiedel, a private-sector archaeologist, wrote a withering dismissal of every single piece of evidence presented.

In his book, Adovasio (who sided with Dillehay on this one) suggested that Fiedel "reserve some space in the State Home for the Terminally Bewildered." Adovasio whacked Fiedel as "a previously little known archaeologist now working for a private salvage archaeology firm" who "has no field experience in Paleo-Indian sites or complex late Pleistocene or Holocene sites" and "has published one rarely used prehistory textbook but otherwise has no apparent credentials."*

Archaeology's caste system is another facet of the discipline that makes it more amateurish a science than, say, particle physics. How many weekend astrophysicists could write up a report challenging Stephen Hawking that would be widely accepted as truth? When new evidence in, say, particle physics opens up a Kuhnian melee, the folks who rush in to the breach tend to be, well, particle physicists. But in

*Professional archaeology is a sort of caste system. The Brahmins are the credentialed, tenured professors at known colleges. They publish in peer-review journals. Beneath them are private-sector archaeologists, also known as salvage archaeologists. They might publish in popular journals such as *National Geographic*, but their day work is something different altogether. They determine for, say, a mall developer whether there are any "significant" remains on a piece of real estate slated to become a food court. Below the salvagers are the rank amateurs and hobbyists who often spend a weekend out at some site hoping to find a Clovis point or two to sell on eBay or to keep in their special cigar box back home.

prehistoric archaeology, with its rather elastic sense of membership ranging from well-credentialed academics like Adovasio to salvage archaeologists to slightly bonkers theorists to ranting neo-Nazis, all of them can rush right in. And do.

What underlies the mudslinging use of bloated Latinisms as well as the compulsion to make a show of tidy whisk brooms and Euclidean grids is the sense, maybe even fear, that archaeology is not a science at all. There's a lot of play in the carbon dating, all the evidence is in dispute, and, sure, maybe the elders' caution can easily be dismissed as a Freudian conflict of interest. All of this means that the pre-Clovis evidence requires a lot of interpretation, a fact that makes it very easy for personal desire and anxiety to leach like groundwater into that drawerful of cobbles and lithics. As one defender of Dillehay confessed in his own report, "I wondered if, by being too close to these stones for too long, I was building an interpretive sandcastle."

But the sandcastle's been built, and some have begun to tell a new creation story—about just who pre-Clovis Man was, where he came from, how he lived and died. The sudden appearance of this yarn explains why prehistoric archaeology really isn't as much a science as it is a form of tribal narrative. These stories have less to do with what's obvious from the evidence than with what some deeply long to hear. It's time to look closely at the story that's getting told right now about the earliest inhabitants of this continent. I have a little experience in this field. I know how to jury-rig a narrative using only a couple of wayward factoids to make it sound just right. It's something I was born to do. I am the direct descendant of Charlemagne.

V. Ladies and Gentlemen, Kennewick Man

FOR MOST OF THE 1990S, the sotto-voce chatter about pre-Clovis Man and his possible identity was little more than politically mischievous buzz out on the edge of archaeology. Insiders talked about spearpoints and some disputed bones, DNA, and cordage, but it wasn't a story so much as it was narrative tinder, very dry, waiting for a spark.

That spark finally flew one hot summer afternoon in 1996, on the banks of the Columbia River. Some college students were trying to sneak into a hydroplane race, and as they stomped through the muck of a bank, one of them saw a few bones and then pieces of a skull.

The find eventually was passed on to a local forensic expert, a salvage archaeologist who worked out of a converted rec room in his house. He would become the rhapsode for these bones. Divinely, his name was James Chatters.

Chatters released the carbon dating that put the bones as far back as 7600 BC. He also described a Cascade point embedded in the hip. This style of Paleo-Indian arrowhead is a long, thin design that would fit right in with the skeleton's age. But Chatters also said that he didn't believe this skeleton belonged to a Paleo-Indian but rather to "a trapper/explorer who'd had difficulties with 'stone-age' peoples during his travels."

In other words, this skeleton represented a crime scene, and the victim was not Paleo-Indian.

Immediately, several Indian tribes, such as the Umatilla, demanded the bones, charging that they had to be of Native-American heritage. Over the next few years, what was at first a strange political dustup grew into an even more bizarre legal battle. The scientists held to simple principles of open inquiry: all we want to do is examine the skeleton more closely. The Native Americans suspected a ruse to get around new laws protecting the burial of ancient bones.

In 1990 President George H. W. Bush signed the Native American Graves Protection and Repatriation Act (NAGPRA), which sought to make amends for the grave robbing and bizarre antics of the previous decades. In the nineteenth century, the Smithsonian Institution wanted Indians' skulls to mount on display. So quite often after a battle, Indian corpses were decapitated, the heads packed in boxes and shipped back to Washington to be "studied." The money was good enough that often violence broke out when Indians saw white men—the emissaries of European civilization—loitering around a burial ground, since the suspicion was that they were waiting around in order to dig up grandma and cut off her head. A few centuries' worth of desecration of the Indian body is something mainstream history still avoids. It's hard for non-Native Americans today to understand all the lingering resentment. Try this on: toward the end of the Civil War, in Denver, a group of marauding white men provided theatergoers with a mid-show display of fresh Indian scalps—not merely from heads but from women's vaginas as well. The audience whooped with approval.

Some estimates put the number of Indian skeletons held in museums at two hundred thousand. NAGPRA was an attempt to return those and make sure it didn't happen again. It decreed that all Native-American remains that could be culturally identified were to be returned to the appropriate tribe.

In the past, a number of skeletons that fed the pre-Clovis rumor mill had in fact been seized and reburied. In the back alley of amateur American archaeology, these are notorious: the 10,800-year-old Buhl Woman found in Idaho in 1989, the 7,800-year-old Hourglass Cave skeleton found in Colorado, the 7,800-year-old Pelican Rapids Woman skull, and the 8,900-year-old Browns Valley Man, both found in Minnesota—all reburied.

The Native Americans in Washington State immediately assumed that this talk of Kennewick not being a Paleo-Indian was little more than a scientific tactic to get around the requirements of NAGPRA. Whatever the merits of the case, the issue quickly got caught up in contemporary politics. At the time of the discovery, the Umatilla Indians were working with the Clinton administration to dispose of some chemical weapons (WMD, as we say nowadays). The federal government desperately wanted tribal support on this difficult matter. By the late 1990s, the Umatilla had a casino, which meant they had political and financial clout and couldn't easily be kicked around. When they screamed for the bones, the Clinton administration jumped. Bruce Babbitt, the then secretary of the interior, ordered that the U.S. Army Corps of Engineers take control of the bones. Then, to "stabilize" the site where the bones were found, the army choppered in five hundred tons of riprap and buried the bank. The archaeological site was protected by being destroyed.

Chatters and the group of scientists who had gathered around him, calling for an open inquiry into the skeleton, were stunned. This wasn't just politics; it was medieval obscurantism. To the scientists, this was the equivalent of locking Galileo in his room and demanding that he recant. And there the story might have ended, but for one word that totally changed the nature of the debate in the pop culture outside the courtroom. When Chatters first examined Kennewick in his rec room, he looked at the skull and then described what he saw as "Caucasoid-like."

The narrative tinder of decades suddenly exploded in flames, and from the fire, like a phoenix, arose a new and wild story: A Caucasoid man, who was among the First Americans, was murdered by genocidal newcomers, Mongoloid invaders coming across Berengia after the last ice age.*

To color the fight with the absurdity peculiar to all truly American events, the Asatru Folk Assembly—a neo-Norse movement that claims to represent the "native European" religion—asserted its rights to Kennewick's bones. The neo-Norsemen argued that they were the nearest tribe related to Kennewick Man and that under NAGPRA they should be given the bones for reburial. The federal courts did not give them Kennewick but did allow them to perform funeral rites over the bones. And so a year after that hydroplane race, big hairy blond men wearing horns and garish furs performed the Norse burial ceremony in Washington State for their mourned errant ancestor.

*Throughout the theories and quarrels surrounding this prehistory, there is a strange kind of recapitulation going on. Every theory propagated about the European conquest of the Indians after Columbus seems to have its Doppelgänger in the pre-Clovis era. Just as American Indians were the victims of genocide in the colonial period, so it seems were the early Caucasoids at the hands of Paleo-Indians. Some theories say that the early Caucasoids were wiped out by germs, a recapitulation of the account of smallpox-infected blankets that has become a near parable in American history. In this way, the Indians' attempt to claim the Kennewick skeleton is simply the evil twin of nineteenth-century grave robbing.

VI. A Brief History of Caucasians

DOES RACE EXIST? Of course it does. We see it every day. Guy steals a purse, the cop asks, What did he look like? You say, He was a six-foot-tall black guy, or a five-and-a-half-foot-tall Asian man, or a white guy with long red hair. As a set of broad descriptions of how people look, race exists.

If you were to look at me, you would easily categorize me as Caucasian. I'm the ruddy sort that burns quickly, with reddish hair now shading into white. Most people hazarding a guess might say Scottish, which is what I have always said. Just to be sure, I recently submitted my DNA to see what the incontrovertible scientific evidence might show. The result was surprising (though in some ways not surprising): I carry the DNA marker found in great abundance among the Fulbe tribe of contemporary Nigeria.

Sure, maybe the marker is about as significant as my Charlemagne genes. On the other hand, that very Nigerian coast is the tribal location where many slaves were captured and held in the notorious slave castles until traders' galleys could transport them to American ports. The main harbor that received more slaves than any other on the eastern seaboard was Charleston, South Carolina. My mother's family has lived there nearly three hundred years. Maybe I have a Thomas Jefferson problem.

Whenever it was that a black woman* entered my bloodline (and my white ancestor entered hers), it's no longer apparent in the way I now look. I am Caucasian as surely as my Fulbe cousins are black, because race is a set of visual cues we all recognize—skin shade, nose shape, eyelid folds, cheekbone prominence, etc. We hold these vague blueprints of race in our heads because, as primates, one of the great tools of consciousness we possess is the ability to observe patterns in nature. It's no surprise that we'd train this talent on ourselves.

*I trust that I don't need to explain why I make this assumption.

Here's another example, a little closer in time. My grandmother was Weinona Strom. Her first cousin was Strom Thurmond, which makes the late senator my first cousin, twice removed. It also makes his half-black daughter, Essie Mae Washington-Williams, my second cousin once removed.

For those of us who have had to contend with Strummy-boo* all our lives, looking at Essie Mae and seeing the senator's face gazing out from her own is a kind of thrilling shock. But what's far more interesting is Essie Mae's daughter. Because Essie Mae married a man our pattern-seeking brains would recognize as black, the evidence of Strummy's whiteness is practically gone only one generation later. I suspect that among the great grandchildren, Strummy's presence in the Washington-Williams family will be as washed away as the Fulbe tribe is in me.

Yet the notion of race as an unchanging constant through time is as old as the Bible. When Noah's Flood receded, the three boys Japheth, Shem, and Ham went out into the world to engender white people, Semites, and all others, respectively. This doesn't quite shake out into the later notions of white, black, and yellow, but you get the idea. The boys are still with us. The early word "Shemitic" settled down to become "semitic." And among amateur chroniclers writing in the ponderous style of the town historian, it's not hard to find references to the "Hamitic race" as a way of saying "black folks." Japheth never became a common adjective, perhaps because of that thicket of consonants. More likely, though, it's because whites appointed themselves the Adamic task of naming the other races. It was not until the Age of Reason that scientists tried to figure out empirically what race meant and how it came to be. The signal year was 1776, with the publication of a book called *On the Natural Varieties of Mankind* by German biologist Johann Friedrich Blumenbach.

At the time, Blumenbach's theory had a certain symmetry that made it the very model of good science. These days, his theory seems

*That's really what we called him.

insane. He argued that Native American Indians were the transitional race that eventually led to Asians. (Don't try to work out the geography of this: it will make your head explode.) And another group—which Blumenbach simply conjured from a far-away people, the "Malayans"—evolved over time to become Africans. (Again, if you're puzzling out the geography, watch your head.)

At the center of all this change was the white race, which was constant. Blumenbach believed darkness was a sign of change from the original. All of mankind had fallen from perfection, but the darker you were, the farther you had fallen. As a result, the best way to locate the original Garden of Eden, according to Blumenbach, was to follow the trail of human . . . *beauty.* The hotter the women, the hunkier the men, the closer you were to what was left of God's first Paradise. Here is Blumenbach explaining the etymology of the new word he hoped to coin:

> I have taken the name of this variety from Mount Caucasus, both because its neighborhood, and especially its southern slope, produces the most beautiful race of men, I mean the Georgian . . .

Blumenbach's theory is totally forgotten today by everybody (except maybe Georgian men). All that remains is a single relic, the word he coined for God's most gorgeous creation—"Caucasian."

The word itself is lovely. Say it: Caucasian. The word flows off the tongue like a stream trickling out of Eden. Its soothing and genteel murmur poses quite a patrician contrast to the field-labor grunts of the hard *g*'s in "Negroid" and "Mongoloid." Caucasian. The exotic isolation of those mountains intimates a biblical narrative. You can almost see it when you say it: the early white forebears walking away from Paradise to trek to Europe and begin the difficult task of creating Western Civilization.

Ever since Blumenbach launched this word two and a half centuries ago, the effort to pin down the exact and scientific meaning of race has never ceased. Even today, the U.S. Census is little more than an explosion of ethnic agony that arrives every ten years like constitutional clockwork. The number of races has expanded and contracted wildly between Blumenbach and now, depending on the mood of the culture. The basic three have gone through scores of revisions, growing as high as Ernst Haeckel's thirty-four different races in 1879 or Paul Topinard's nineteen in 1885 or Stanley Garn's nine in 1971. Today, we nervously ask if you're white, African American, Native American, Asian, or of Hawaiian or Pacific Islander descent.

But it wasn't that long ago that the question would have turned upon races only our great-grandfathers would recognize. Let us mourn their passing: the Armenoids, the Assyroids, the Veddoids, the Orientalids, the Australoids, the Dalo-Nordic, the Fälish, the Alpines, the Dinarics, the Fenno-Nordic, the Osteuropids, the Lapponoids, the Osterdals, the Cappadocians, the Danubians, the Ladogans, the Trondelagens, and the Pile Dwellers.

In the meantime, science has made its discoveries. The mystery of race has been solved. For the longest time, scientists were stymied by a contradiction. Surely skin tone had something to do with colder climates creating paler shades, but then why weren't Siberians as pale as Swedes, and why were Eskimos as dark as equatorial islanders? The answer was announced in 2000, but it's so tedious hardly anyone noticed.

Skin pigmentation changed long ago not only to protect skin from different levels of sun exposure—that's obvious—but also in order to regulate the amount of vitamin D_3 manufactured by the sun just under the skin. This is the theory of Professor Nina Jablonski, a paleoanthropologist with the California Academy of Sciences. So when the first swarthy inhabitants of modern Scandinavia confronted a lack of ultraviolet light, their kind quickly selected out for paler children whose skin would manufacture enough vitamin D_3 to keep them healthy. Meanwhile, Eskimos arrived in the Arctic dark-skinned. The local cuisine of seal and whale is rich in vitamin D_3, so the skin was never summoned into action. Evolution has one big rule. If there's no pressure on the system to change, then it doesn't bother. So Eskimos remained dark.

When we look at the different races, according to Jablonski's theory, what we're actually seeing is not "superiority" or "good people" or "race." All that we are seeing, the *only* thing we are seeing when we look at skin color, is a meandering trail of vitamin D_3 adaptation rates.

VII. The Mounting Evidence

SCIENCE PREFERS TO CONFIRM its newest findings with the newest tools. Just as fingerprinting is no longer the gold standard of guilt or innocence now that DNA testing is the rage, archaeologists have a few new tricks. These cutting-edge techniques come with gleaming names—Optically Stimulated Luminescence, Electron Spin Resonance Dating, and Accelerator Mass Spectrometry—and they are confirming pre-Clovis dates in ways that make carbon dating look like counting tree rings. By the time we figure out how these techniques are flawed, of course, our prejudices will be so well muddled among the tentative facts that they will be as inextricable as ink from milk.

According to the revolutionaries heralding pre-Clovis Man, that hardly matters, since so much other corroborative evidence is appearing. Some lab tests reveal that Native American Indians apparently have a signature strand of DNA known as Haplogroup X. The only other large population on the earth carrying this genetic marker is Europeans. The suggestion is that there must have been intermarriage in North America before Columbus, possibly before the last ice age. Moreover, now that the Iron Curtain has fallen, archaeologists have been able to do more digging in Siberia, where they expected to find Clovis points or something like them. They haven't. This absence, as well as the presence of Haplogroup X, has led some people to theorize that although Clovis Man might have crossed over to North America thirteen thousand years ago at the end of the last ice age, he would have encountered people already here—people possessed of the X gene as well as the Clovis tool kit.

Who might these people have been, and where might they have come from? One prominent theorist with an answer is America's chief archaeologist at the Smithsonian. A big, bearded bear of a man, Dennis Stanford could pass for a Norse king in some other time. Stanford has struggled with the mystery of why Clovis points *don't* show up in Siberia. He notes that they resemble the early work of Solutrean culture. The Solutreans were prehistoric people who lived in modern-day France and Spain some eighteen thousand years ago. They are perhaps most famous for being the possible candidates for painting the horses of Lascaux and their own hands on the walls of the Altamira Cave. Stanford argues that their tool kit, which included stone points, looks like a predecessor to the Clovis style.

"There must be fifty or sixty points of comparison," he has said.

He believes that these proto-Europeans must have been intelligent enough to make watercraft. Hugging the coast of what would have been a glacier all around the crescent edge of the northern Atlantic, they sailed away to a new land.

Other scientists are providing even more evidence that seems to support these general ideas. Several anthropologists have daringly revived the argument that examining skull shapes can reveal ethnicity. Douglas Owsley, also now at the Smithsonian, and his partner, Richard Jantz, at the University of Tennessee, have put together collections of measurements, described by *Newsweek* as a database of "2,000 or so profiles that consists of some 90 skull measurements, such as distance between the eyes, that indicate ancestry." They have developed software that allows them to input a bone's measurements and receive "ethnicity" as an output.

Among their fans and followers, there is talk of some of the peculiar skeletons found over the years. A 9,200-year-old body known as Wizards Beach Man, found at Pyramid Lake, Nevada, in 1978, was determined to be possibly of "Norse" extraction and to have "no close resemblance to modern Native Americans." Another skeleton, known as Spirit Cave Man, was found in Nevada in 1940. His bones date to 7450 BC. When his skull measurements were run through the software, out spat a finding of "Archaic Caucasoid."

Once again, there's Blumenbach's word. Only this time it's got that "oid" ending. What is the difference between Caucasoid and Caucasian?

"Cauca*soid* sounds more scientific," University of North Carolina anthropologist Jonathan Marks told me, laughing. Otherwise it has no more meaning or significance than Blumenbach's original. Caucasoid is a magnificent piece of pure Star Trekkery, a word meant to sound all clinical and precise, even nerdy. But the word is a rhetorical Trojan horse. Its surface meaning suggests something scientific, respectable, and learned, when in fact what we really hear are the connotations lurking inside, long-suppressed intimations of superiority, exceptionalism, and beauty.

VIII. Kennewick's Biography

The court fight over Kennewick Man was resolved this January in favor of the scientists—in part because this is America and who can be against "open inquiry"? Yet the ramifications for Native Americans and for white Americans will be immense. In the popular market of ideas, the decision by the courts affirms a lot more than the noble virtue of open inquiry. It legitimizes the *story*—the story of the Caucasian man who came to this continent as the Authentic First American and whose bones survived the millennia to report the truth.

And the story that has been told these last eight years about this hundred-century-old man is marvelous in its perverse beauty. It begins with his name. Does anything sound more European, more positively British, than Kennewick? Native Americans had dubbed him the "Ancient One," but it didn't take. The mass media, which follows the meandering will of the popular mob, could sense where this story was trending, and so they ran with "Kennewick." Isn't that a suburb of Essex, or London's other airport? Perhaps not so ironically, the name is an anglicized version of the Indian word "kin-i-wak," meaning "grassy."*

In the few years after Sir Kennewick's discovery, his life was described and depicted in all the leading magazines. One writer on the subject, Sasha Nemecek, confessed that when she looks at the evidence "the misty images of primitive explorers evaporate" and now "I suddenly picture a single artisan spending hours, perhaps days, crafting these stone tools" whose "workmanship is exquisite, even to my untrained eye." To accompany her article, an artist rendered images of what Kennewick's ilk looked like. You might mistake him for an English professor at Bennington, but in fact he's the First American.

And his bride has the complex tool kit of her time, not to mention a nice Ann Tayloroid dress and a haircut that presages Jennifer Aniston by nearly ten millennia. She has thoughtfully shaved her legs for the artist, the better to see her lovely Caucasian skin.

Where did these pictures appear? *Scientific American.*

Right away, Kennewick was described with words that launched him millennia ahead of his primitive enemies, the Paleo-Indians. He was, as Chatters had said, probably an individual "trapper/explorer"—two words that, together, imply degrees of complex thought far in advance of his time, especially when set up against a mass of "stone age peoples." An article in the local newspaper, the *Tri-City Herald,* painted beautiful scenes of Kennewick as the "strongest hunter in his band." Paleo-Indians were still mucking around in "tribes" while Kennewick traveled with his "immediate and extended family members."

Food was important. "To keep up his strength, he and his band dined on rich, lean roasts and steaks . . ." Kennewick is, naturally, on the Atkins diet. No Type II diabetes for Kennewick.

*Just when Kennewick was discovered, another ancient skeleton was found on Alaska's Prince of Wales Island. This skeleton was quickly declared to be "Prince of Wales Island Man," making it seem as if the ancient forebears to the Saxon kings thought of the Pacific Northwest as a dandy vacation spot.

Kennewick Man received glamour treatment from all the major media, in which he was lauded for his near modernity. Lesley Stahl's piece on *60 Minutes* introduced Kennewick to television viewers as someone with "a tremendous amount of symmetry to his body" and therefore "handsome." (Blumenbach's notion of superiority as beauty is never really behind us.) Stahl permitted Chatters to say that Kennewick possessed "a lot of poise." The *Washington Post Magazine* took note of Kennewick's "ambition."

In a *New Yorker* article, we learn that "Some nearby sites contain large numbers of fine bone needles, indicating that a lot of delicate sewing was going on." The needles might have belonged to the Paleo-Indians, or: "Kennewick Man may have worn tailored clothing."

Swish that word around on your connotative palate. *Tailored.* Feel the force, tugging us in a certain direction.

Newsweek's cover story noted that skulls like Kennewick's are so different from what archaeologists expect that they "stand out like pale-skinned, redheaded cousins at a family reunion of olive-skinned brunettes."*

In these stories the Indians are typically ignored or they simply

*And these are the elegant accounts that struggle to keep the story contained inside the scientists' own cautious terms. From there the implications of Kennewick quickly became insinuated in current fashions of political opinion. John J. Miller of the *National Review* mentions a "growing suspicion among physical anthropologists, archaeologists, and even geneticists that some of the first people who settled in the New World were Europeans." Note how a tentative resemblance of skull shape, "Caucasoid-like"—always hedged by the scientists—has quickly settled into declarative certainty: "were Europeans." The politically obvious conclusion is also clear, as the writer continues: "An important part of American Indian identity relies on the belief that, in some fundamental way, they were here first. They are indigenous, they are Native, and they make an important moral claim on the national conscience for this very reason. Yet if some population came before them—perhaps a group their own ancestors wiped out through war and disease, in an eerily reversed foreshadowing of the contact Columbus introduced—then a vital piece of their mythologizing suffers a serious blow." Once you step away from the magazines and books, the story drifts into the poisonous domain of the Internet, where discussions tend toward a brutalist reduction, like this comment from shmogie1 on the alt.soc.history board: "Kennewick man is older than any known N/A [Native American] remains, and appears to be much more European than N/A, so your people stole the land from my European ancestors who were here first."

move about as a supernumerary horde brought onstage to throw the Cascade point and bring down the handsome Kennewick with his poise and ambition and all the other adjectives that will eventually lead to the abandonment of nomadism, the invention of agriculture, and on to the foundation of society that would lead us inexorably toward Western Civilization.

Which, in turn, would bring Kennewick's Caucasoid-like descendants back to America to find him and tell his story.

IX. Kennewick's Back Story

THE STORY OF AN EARLY European presence here in America would be fascinating if it hadn't already been told so many times. The number of theories holding out Native Americans as either latecomers or Europeans in loincloths is endless. Even the earliest depictions of Indians simply used European bodies and faces with a few feathers added.

Western Europeans were stunned that the New World had so many people already in it. How could these primitives have gotten there first? They must be . . . us! Theories abounded. Some in England thought the Indians were covert Welsh families who'd slipped over on their rafts. Others wondered whether they weren't the lost people of Atlantis. A whole host of arguments had it that Indians were Jews. During the colonial era, the chief rabbi of Holland, Menashe ben Israel, claimed that all Native Americans were descended from the Lost Tribes of Israel, and the theory was confirmed by a 1650 book entitled *Iewes in America, or, Probabilities that the Americans are of that Race.* Mormons continue to believe this account of Native American origin, holding that the sons of Lehi sailed to the Americas around 600 BC and lost their traditions and knowledge of the Torah. They reverted to a state of savagery, and their descendants scattered among the plains and throughout the two continents. Thus all Indians are, essentially, Jews Gone Wild.

Most Americans rarely saw images out of books such as the rabbi's.

Rather, the most widely available image was to be found on the coins in your pocket. European, if not Roman imperial.

In 1914, on a ten-dollar-gold coin, it was still possible to see in the face of the Indian the wavy-blonde Nordic princess of our dreams. Practically at the same time, in 1913, an image that registered in our pattern-constructing brains as "Indian" would finally appear on the famous Buffalo nickel.

It's important to know this history and tradition when you consider the image conjured by Chatters when he asked an artist to take the Kennewick skull and reconstruct the face. Well, if only that was precisely what he did. But Chatters didn't just hand the skull to someone and ask him to reconstruct the face. Instead, he had an epiphany, as he explained once, right at home: "I turned on the TV, and there was Patrick Stewart—Captain Picard, of *Star Trek*—and I said, 'My God, there he is! Kennewick Man.' "

Forensic reconstruction is a very iffy "science." The problem is that the features we look to for identification are fleshy ones—ears, nose, and eyes—and those are the most difficult to know from a skull.* So reconstruction is more art than science, or, with its stated success rate of roughly 50 percent, about as good a predictor as a coin toss. Consider what Chatters did: by making Kennewick Man perfectly resemble one of the most famous pop-culture Brits of our time, he allows the visual cues to confirm his finding and so avoids even the need to repeat the word "Caucasoid-like."

Kennewick's skull is often described as "narrow, with a prominent nose, an upper jaw that juts out slightly, and a long narrow braincase," or, more properly, dolichocranic and slightly prognathous, marked by a lack of an inferior zygomatic projection. Yet here's the problem with

*I and every other writer call Kennewick's head a "skull." The implication is that it was found whole. In fact, it was found in parts that Chatters pieced together. When government experts put the pieces together, they built a skull whose dimensional differences from Chatters's version were deemed statistically significant. Again, at every stage of this story, the details get pushed toward the Caucasoid-like conclusion.

looking at those vague features and declaring them "Caucasoid." We don't really know what people's skulls looked like ten thousand years ago. We have only a few, like the pre-Clovis points, so it's reckless to draw any conclusions. Skull shapes, like skin color, can change more quickly than we think, especially if there has been traumatic environmental change.

Franz Boas, the legendary anthropologist from the turn of the last century, debunked a lot of skull science in his time by proving that the skulls of immigrant children from all parts of the world more closely resemble one another than do their parents'. Rapid dietary shifts can cause major structural changes in skeletons—just ask the average Japanese citizen, who has shot up four and a half inches in height since World War II, or the average American man, who has packed on an extra twenty-five pounds since 1960. The truth is that there exists no coherent history of skull shapes back through time, so to say that a ten-thousand-year-old skull resembles a modern white-guy skull is to compare apples and oranges.

In time, Chatters tried to calm the storm of his unscientifically absurd remarks. He repeatedly said things like this: Kennewick Man "could also pass for my father-in-law, who happens to be Scandinavian." Then one day he was suddenly insisting, "Nobody's talking about white here."*

He insisted that he meant that the skull simply didn't resemble the classic "Mongoloid" features of Asia. He said that Kennewick could have been Polynesian or even ancient Japanese. It turns out that those vague Caucasoid features are also found in the Ainu people of prehistoric Japan, as well as in other non-European peoples.

Don't be confused here. The scientists themselves who fling around words like "Caucasoid" are the very ones who also admit that

*His contradictions are maddening. At one point, Chatters said: "I referred to the remains as Caucasoid-like . . . I did not state, nor did I intend to imply, once the skeleton's age became known, that he was a member of a European group." But then he told Elaine Dewar, the author of *Bones: Discovering the First Americans*: "I say you can say European. Who can prove you wrong?"

the "Caucasian" skull is found everywhere. That's right. This Caucasian skull shape is found all over the planet. For example, another ancient skull always brought up alongside Kennewick's is a female skull found in Brazil. Nicknamed Luzia, the skull was analyzed in a report that cited the following locations for resemblance: skulls seen among early Australians, bones found in China's Zhoukoudien Upper Cave, and a set of African remains known as Taforalt 18. So we've narrowed it down to Australia, China, and Africa.

Another study of an ancient skeleton known as Spirit Cave Man narrowed down his skull-shape origin to: Asian/Pacific, the Zulu of Africa, the Ainu of Japan, the Norse, or the Zalawar of Hungary.

What conclusion can be drawn from finding Caucasian skulls in Asia? Or finding African skulls in Brazil? Or finding Polynesian skulls at the continental divide? Is it that these "groups" traveled a lot, or that skull shapes change radically and quickly over time? It's the latter, of course, and plenty of anthropologists have known that for some time. In the early twentieth century, Harvard anthropologist Earnest Hooten documented the wide variety of skull shapes he found among ancient Native Americans.

As Jonathan Marks explained it to me, Hooten "studied Native American skulls from precontact all the way to the eighteenth century, and he sorted them into cranial racial categories. He called them 'pseudo-Australoid' and 'pseudo-Negroid' and 'pseudo-Mediterranean' because they had those features. He was smart enough not to say, 'Well, I guess these people encountered a stray Australian aborigine on his way to Colorado.' Clearly he recognized that there was considerably more diversity in early Indian skulls than he was used to seeing."

What this suggests is not so much that Africans, Mongoloids, and Europeans were storming the American shores ten thousand years ago but rather that in any one group, at any one time, you will find all sorts of anomalies.

The Center for the Study of the First Americans, at Texas A&M University, is a clearinghouse for pro-Caucasian theories of early

America. The center publishes a manly newsletter, "Mammoth Trumpet," in which one can find a set of arguments that inspire a kind of sorrow and pity. The founder, Dr. Robson Bonnichsen (like so many of these academics: the look of a Norse king with a big bushy beard), was commonly quoted stating things like this: "We're getting some hints from people working with genetic data that these earliest populations might have some shared genetic characteristics with latter-day European populations." Maybe he doesn't know that he's the direct heir to King Charlemagne.

What makes the claim all the more paltry is that once you start reading about the European connection to pre-Clovis Man here in America, you can't help but notice that the same essential story is getting told in other, completely separate fields—such as when the first Europeans evolved or when early ape creatures crossed over the line leading to humans. All of them make claims that have the contours of the same fight—the revolutionaries challenging the traditionalists, all of them finding a way to shoehorn Europeans into a story with hints of superiority and beauty.

The current theory about the beginning of mankind—the Out of Africa theory—states that an early prehuman, *Homo erectus,* evolved into *Homo sapiens,* who then left Africa some one hundred thousand years ago and eventually evolved into the modern peoples of the world. But there is a small contingent of rebel theorists—the "multiregionalists"—who hold that it was *Homo erectus* who spread out to various locations where each developed into its own transitional hominid. In Asia: Peking Man. In South Asia: Java Man. And in Europe: Neanderthal Man. Each of these specimens would eventually evolve simultaneously into *Homo sapiens.* According to the rebels, there was some gene mixing at the margins of these separately developing species to keep the general hominid ability to reproduce together. It's a serviceable theory that manages to keep all mankind barely in the same species while creating an intellectual space for racial differences and European uniqueness. It is the "separate but equal" theory of physical anthropology.

As theories go, multiregionalism can be pretty slippery, but then it has to be. New evidence constantly confounds. Not long ago, DNA tests revealed that Neanderthal made no direct genetic contribution to modern man. So multiregionalists now struggle to keep Neanderthal in the picture at all, arguing that there was some sex among the different humans and that the evidence is with us. One of the arguments is that my big nose (as well as those great beaks on Jews and Arabs) is telling evidence of Neanderthal genes. That's the theory of Dr. Colin Groves, a very dolichocranic man who, it should be no surprise by now, sports a big beard.

There are so many theories in which the key moment of development that "makes us human" somehow occurred in Europe that I have begun to collect them, like baseball cards. Those cave paintings in Lascaux and Altamira, for instance, are often held up as the threshold event revealing "abstract" thought, which made us truly human. My personal favorite, this week, takes us all the way back to the apes. A few years ago, David Begun, a primate specialist in Toronto, announced that he'd found our last common ancestor with the great apes; i.e., the notorious missing link. Where? In Europe. His theory holds that African apes crossed into Europe, picked up those civilizing traits that would eventually lead to humanness, and then slipped back to the dark continent just under the deadline for their Out of Africa journey. Scientists are now finding these apes all over Europe. Just last winter another one was excavated near Barcelona and heralded as further proof of Begun's theory. The researchers remained tight-lipped about what it all meant, but popular outlets found ways to get the point across, such as this sentence in a recent CBS News report: "The researchers sidestepped a controversy raging through the field by not claiming their find moves great-ape evolution—and the emergence of humans—from Africa to Europe."

X. A Caucasian Homecoming

The question of just when we became human gets answered in our popular press all the time. Was it when we assembled the first rudimentary tool kit, or when we grunted out the earliest phonemes of complex language? Was it when we made those paintings in Altamira and Lascaux, or when we left off being knuckle-dragging ape-like critters and stood up? Standing up has been a particularly fertile field for this kind of musing, with theories ranging from cooling off to intimidating other species to freeing the hands. I'd always heard that we abandoned squatting because we wanted to see over the top of the grass on the African savannahs, allegedly our first habitat. One early 1980s theory was that standing evolved for "phallic display directed at females."*

Last year a British scholar named Jonathan Kingdon argued in his book, *Lowly Origin,* that our standing up probably had a lot to do with getting food and happened in undramatic stages, first by straightening the back while squatting and later by extending the legs—all of this happening in tiny incremental stages over vast swaths of time.

As theories go, that's not nearly as much fun as "seeing over the grass" or "phallic display," but it has the ring of truth to it, a ring that, let's face it, will never endear such an idea to writers of newsweekly cover lines. Which is also why you've never heard of Jonathan Kingdon.

Scientists like to invoke Occam's razor, the principle that the simplest explanation is often the most truthful. These days we have almost the opposite problem: pop thinkers tend to oversimplify in a way meant to attract attention. The first time I ever got a whiff of this was when I was a teenager reading Desmond Morris's book *The Naked Ape.* Morris theorized that the reason human females had big breasts (as opposed to the tiny sagging dugs of other primates) was because we had discovered love. In doing so we switched from copulating doggie style to the more romantic missionary position. But all those eons of looking at the round globes of the female's buttocks from behind had developed into the image stimulus required for the maintenance of erections during intercourse. Therefore, Morris argued, the human

*Were this the case, every animal in nature, down to the amoeba, would stand.

male still needed large rounded visual cues and, according to the rules of Darwin, was rewarded with great big hooters.

Even as a kid, I remember thinking, excellent, but really? Morris's simplicity makes monstrous assumptions that just so happen to yield a theory pre-edited for the short punchy demands of modern mass media. A hook, if you will. (Not that it didn't work: thirty years after reading that book, the only detail I can remember is the boob theory.) Morris's theory has little to do with truth and everything to do with selling books. Perhaps it's time to set aside Occam's razor and pick up Morris's razor, which shuns any theory that might excite a cable-television producer and elevates the plodding theory that makes a kind of dull, honest sense.

Apply Morris's razor to Kennewick Man and here's what you might get:

Chances are that Adovasio and his colleagues are right about the basic assertion of an ancient arrival of *Homo sapiens* to this continent. It easily fits in with what else is known. For instance, the archaeological record in Australia is redundant with evidence that aboriginals arrived there at least fifty thousand years ago. That journey would have required boating some eighty miles, many believe. So it's perfectly conceivable that there were multiple entries to the American continent, with at least one crew, probably Asians like the Ainu, lugging their Haplogroup X gene and following the food (not "exploring") by canoe or on foot across the Berengia bridge, possibly just after the penultimate ice age, circa twenty thousand to thirty thousand years ago, giving them plenty of time to leave some pre-Clovis fossils.

That's one story, a very Kingdon-like theory, all very probable but not a very good cable special or science-magazine cover story. Morris's razor, though, spares us the rest of the theory, according to which the First American is of an ancient tribe whose members just happen to resemble the very scientists making the claim and whose sad end came about after a genocidal campaign against these superior but outnumbered Caucasoids by hordes of Mongoloid stone age peoples.

This epic extrapolation is drawn from one single Cascade spear-point—a leap about as likely as a Martian anthropologist staring at an Enfield bullet, a scrap of gray wool, and a dinged canteen, and then successfully imagining the states'-rights debate leading up to the nation's Civil War.

The same Martian anthropologist might also quarrel with the pre-Clovisites' view that the Kennewick battle is a latter-day clash between science and religion—the Indians with their mythic stories of origin and the scientists with their lithics and their scientific dates. Given the scant evidence for either, it's more accurate to see the debate as between two forms of folklore squaring off over control of our continents' creation story. In an editorial last year, the *Seattle Times* captured one side of this fight perfectly. Kennewick, the paper said, had "held onto his secrets for more than nine thousand years and now, finally, scientists will get a chance to be his voice."

Why assume the scientists' narrative in this case is closer to the empirical truth? There have been times in the history of archaeology when one could find more objective, hard factual truth in the local oral narratives than in the scientists' analysis, and this may well be one of those times. Oral legends, we increasingly learn, are often based on real events, and those myths can sometimes be decoded to reveal the nuggets of ancient journalistic truth that originally set them into play.

How do we know that the Vikings made a landing at L'anse Aux Meadows in Newfoundland? Because an obsessed lawyer named Helge Ingstad insisted that the Icelandic sagas, the oral epic poetry of his people, were based on fact. No one disputes that the *Iliad* is based on a real war, Ingstad argued, or that the *Song of Roland* derives from an actual tactical blunder by Charlemagne. This small-town lawyer analyzed the details given in the myths and spent years trying to locate the campsite for "Vinland." In 1961 he found it and overthrew the old European story about who arrived first to this continent.

There are several Indian creation stories about coming out of ice. The Paiute tell one that ends this way:

> Ice had formed ahead of them, and it reached all the way to the sky. The people could not cross it. . . . A Raven flew up and struck the ice and cracked it. Coyote said, "These small people can't get across the ice." Another Raven flew up again and cracked the ice again. Coyote said, "Try again, try again." Raven flew up again and broke the ice. The people ran across.

Such accounts are myths, yes, but many Native-American origin accounts involve coming out of ice, which certainly fits into all the theories of America's human origins. So why aren't these stories studied the way Ingstad examined his own sagas? Why is the benefit of the doubt given to the scientists' story? It's quite possible that not a single fact in this new pre-Clovis story is true.

Part of the problem of reading either of these stories is that we no longer have a capacity to appreciate the real power of myth. Most of us are reared to think of myth as an anthology of dead stories of some long-ago culture: Edith Hamilton making bedtime stories out of Greek myths; Richard Wagner making art out of Norse myth; fundamentalist Christians making trouble out of Scripture.

When we read ancient stories or founding epics, we forget that the original audience who heard these accounts did not differentiate between mythic and fact-based storytelling. Nor did these stories have authors, as we conceive of them. Stories arose from the collective culture, accrued a kind of truth over time.

Today we've split storytelling into two modes—fiction and nonfiction. And we've split our reading that way as well. The idea of the lone author writing "truth" has completely vanquished the other side of storytelling—the collectively conjured account. I think we still have these accounts, but we just don't recognize them for what they are. Tiny anxieties show up as urban legends and the like. In the late 1980s, when the queasily mortal idea of organ donation was infiltrating the social mainstream, suddenly one heard an authorless story of a man waking up in a Times Square hotel room after a night of partying to find a stitched wound on his lower back and his kidney missing.

In many ways, the occasional journalistic scandal stems from this tension between the individual as author and the audience as author. The most recent case was *USA Today*'s Jack Kelley. His made-up stories are pure collective desire—stories that we, not just he, wanted to hear. He told the story of the little terrorist boy pointing at the Sears Tower and saying, "This one is mine." Perfect story, finely tuned: The corruption of innocence. American icon as target. The anxiety that terrorism has no end.

Enduring myth can be based on fact, as in Ingstad's case. But often the collective account needs no factual basis, just a mild apprehension that the world is not quite what it seems. No one has ever found a razor blade in an apple at Halloween, nor has any doctor treated anyone for gerbiling.

The story of the Ancient European One is this kind of story, toggling back and forth between the world of fiction and (possibly) nonfiction, authored by a few curious facts and the collective anxiety of the majority.

Because we no longer read mythological stories, we no longer appreciate their immense power. We find ourselves stunned at how something so many deeply long to be true will simply assemble itself into fact right before our eyes. If the majority profoundly longs to believe that men of Caucasoid extraction toured here sixteen thousand years ago in Savile Row suits, ate gourmet cuisine, and explored the Pacific Northwest with their intact pre-Christianized families until the marauding horde of war-whooping Mongoloid injuns came descending pell-mell from their tribal haunts to drive Cascade points into European hips until they fell, one after another, in the earliest and most pitiful campaign of ethnic cleansing, then that is what science will painstakingly confirm, that is what the high courts will evenhandedly affirm, and that is what in time the majority will happily come to believe.

Paul Bloom

Is God an Accident?

FROM THE *ATLANTIC MONTHLY*

There has been of late a vigorous attempt to examine the origins of the human tendency toward religion. A number of explanations have been given: religion as an evolutionary adaptation; religion as a harmonizing cultural artifact; religion as a self-replicating unit of thought, or "meme." Paul Bloom's research has raised another possibility: that religion is the outcome of the dualism inherent in our understanding of ourselves and our world—accidents that arise from our human nature.

I. God Is Not Dead

WHEN I was a teenager my rabbi believed that the Lubavitcher Rebbe, who was living in Crown Heights, Brooklyn, was the Messiah, and that the world was soon to end. He believed that the earth was a few thousand years old, and that the fossil record was a consequence of the Great Flood. He could describe the afterlife, and was able to answer adolescent questions about the fate of Hitler's soul.

My rabbi was no crackpot; he was an intelligent and amiable man, a teacher and a scholar. But he held views that struck me as strange, even disturbing. Like many secular people, I am comfortable with religion as a source of spirituality and transcendence, tolerance and love, charity and good works. Who can object to the faith of Martin Luther King Jr. or the Dalai Lama—at least as long as that faith grounds moral positions one already accepts? I am uncomfortable, however, with religion when it makes claims about the natural world, let alone a world beyond nature. It is easy for those of us who reject supernatural beliefs to agree with Stephen Jay Gould that the best way to accord dignity and respect to both science and religion is to recognize that they apply to "non-overlapping magisteria"; science gets the realm of facts, religion the realm of values.

For better or worse, though, religion is much more than a set of ethical principles or a vague sense of transcendence. The anthropologist Edward Tylor got it right in 1871, when he noted that the "minimum definition of religion" is a belief in spiritual beings, in the supernatural. My rabbi's specific claims were a minority view in the culture in which I was raised, but those *sorts* of views—about the creation of the universe, the end of the world, the fates of souls—define religion as billions of people understand and practice it.

The United States is a poster child for supernatural belief. Just about everyone in this country—96 percent in one poll—believes in God. Well over half of Americans believe in miracles, the devil, and angels. Most believe in an afterlife—and not just in the mushy sense that we will live on in the memories of other people, or in our good deeds; when asked for details, most Americans say they believe that after death they will actually reunite with relatives and get to meet God. Woody Allen once said, "I don't want to achieve immortality through my work. I want to achieve it through not dying." Most Americans have precisely this expectation.

But America is an anomaly, isn't it? These statistics are sometimes taken as yet another indication of how much this country differs from, for instance, France and Germany, where secularism holds greater sway. Americans are fundamentalists, the claim goes, isolated from the intellectual progress made by the rest of the world.

There are two things wrong with this conclusion. First, even if a gap between America and Europe exists, it is not the United States that is idiosyncratic. After all, the rest of the world—Asia, Africa, the Middle East—is not exactly filled with hard-core atheists. If one is to talk about exceptionalism, it applies to Europe, not the United States.

Second, the religious divide between Americans and Europeans may be smaller than we think. The sociologists Rodney Stark, of Baylor University, and Roger Finke, of Pennsylvania State University, write that the big difference has to do with church attendance, which really is much lower in Europe. (Building on the work of the Chicago-based sociologist and priest Andrew Greeley, they argue that this is because the United States has a rigorously free religious market, in which churches actively vie for parishioners and constantly improve their product, whereas European churches are often under state control and, like many government monopolies, have become inefficient.) Most polls from European countries show that a majority of their people are believers. Consider Iceland. To judge by rates of churchgoing, Iceland is the most secular country on earth, with a pathetic 2 percent weekly attendance. But four out of five Icelanders say that they pray, and the same proportion believe in life after death.

In the United States some liberal scholars posit a different sort of exceptionalism, arguing that belief in the supernatural is found mostly in Christian conservatives—those infamously described by the *Washington Post* reporter Michael Weisskopf in 1993 as "largely poor, uneducated, and easy to command." Many people saw the 2004 presidential election as pitting Americans who are religious against those who are not.

An article by Steven Waldman in the online magazine *Slate* provides some perspective on the divide:

> As you may already know, one of America's two political parties is extremely religious. Sixty-one percent of this party's voters say they pray daily or more often. An astounding 92 percent of them believe in life after death. And there's a hard-core subgroup in this party of superreligious Christian zealots. Very conservative on gay marriage,

half of the members of this subgroup believe Bush uses too *little* religious rhetoric, and 51 percent of them believe God gave Israel to the Jews and that its existence fulfills the prophecy about the second coming of Jesus.

The group that Waldman is talking about is Democrats; the hardcore subgroup is African-American Democrats.

Finally, consider scientists. They are less likely than nonscientists to be religious—but not by a huge amount. A 1996 poll asked scientists whether they believed in God, and the pollsters set the bar high—no mealy-mouthed evasions such as "I believe in the totality of all that exists" or "in what is beautiful and unknown"; rather, they insisted on a real biblical God, one believers could pray to and actually get an answer from. About 40 percent of scientists said yes to a belief in this kind of God—about the same percentage found in a similar poll in 1916. Only when we look at the most elite scientists—members of the National Academy of Sciences—do we find a strong majority of atheists and agnostics.

These facts are an embarrassment for those who see supernatural beliefs as a cultural anachronism, soon to be eroded by scientific discoveries and the spread of cosmopolitan values. They require a new theory of why we are religious—one that draws on research in evolutionary biology, cognitive neuroscience, and developmental psychology.

II. Opiates and Fraternities

One traditional approach to the origin of religious belief begins with the observation that it is difficult to be a person. There is evil all around; everyone we love will die; and soon we ourselves will die—either slowly and probably unpleasantly or quickly and probably unpleasantly. For all but a pampered and lucky few life really is nasty, brutish, and short. And if our lives have some greater meaning, it is hardly obvious.

So perhaps, as Marx suggested, we have adopted religion as an opiate, to soothe the pain of existence. As the philosopher Susanne K. Langer has put it, man "cannot deal with Chaos"; supernatural beliefs solve the problem of this chaos by providing meaning. We are not mere things; we are lovingly crafted by God, and serve his purposes. Religion tells us that this is a just world, in which the good will be rewarded and the evil punished. Most of all, it addresses our fear of death. Freud summed it all up by describing a "three-fold task" for religious beliefs: "they must exorcise the terrors of nature, they must reconcile men to the cruelty of Fate, particularly as it is shown in death, and they must compensate them for the sufferings and privations which a civilized life in common has imposed on them."

Religions can sometimes do all these things, and it would be unrealistic to deny that this partly explains their existence. Indeed, sometimes theologians use the foregoing arguments to make a case for why we should believe: if one wishes for purpose, meaning, and eternal life, there is nowhere to go but toward God.

One problem with this view is that, as the cognitive scientist Steven Pinker reminds us, we don't typically get solace from propositions that we don't already believe to be true. Hungry people don't cheer themselves up by believing that they just had a large meal. Heaven is a reassuring notion only insofar as people believe such a place exists; it is this belief that an adequate theory of religion has to explain in the first place.

Also, the religion-as-opiate theory fits best with the monotheistic religions most familiar to us. But what about those people (many of the religious people in the world) who do not believe in an all-wise and just God? Every society believes in spiritual beings, but they are often stupid or malevolent. Many religions simply don't deal with metaphysical or teleological questions; gods and ancestor spirits are called upon only to help cope with such mundane problems as how to prepare food and what to do with a corpse—not to elucidate the Meaning of It All. As for the reassurance of heaven, justice, or salvation, again, it exists in some religions but by no means all. (In fact, even those religions we are most familiar with are not always reassuring. I know some older Christians who were made miserable as children by worries about eternal damnation; the prospect of oblivion would have been far preferable.) So the opiate theory is ultimately an unsatisfying explanation for the existence of religion.

The major alternative theory is social: religion brings people together, giving them an edge over those who lack this social glue. Sometimes this argument is presented in cultural terms, and sometimes it is seen from an evolutionary perspective: survival of the fittest working at the level not of the gene or the individual but of the social group. In either case the claim is that religion thrives because groups that have it outgrow and outlast those that do not.

In this conception religion is a fraternity, and the analogy runs deep. Just as fraternities used to paddle freshmen on the rear end to instill loyalty and commitment, religions have painful initiation rites—for example, snipping off part of the penis. Also, certain puzzling features of many religions, such as dietary restrictions and distinctive dress, make perfect sense once they are viewed as tools to ensure group solidarity.

The fraternity theory also explains why religions are so harsh toward those who do not share the faith, reserving particular ire for apostates. This is clear in the Old Testament, in which "a jealous God" issues commands such as

> Should your brother, your mother's son, or your son or your daughter or the wife of your bosom or your companion who is like your own self incite you in secret, saying "Let us go and worship other gods" . . . you shall surely kill him. Your hand shall be against him first to put him to death and the hand of all the people last. And you shall stone him and he shall die, for he sought to thrust you away from the LORD your God who brought you out of the land of Egypt, from the house of slaves.
>
> —Deuteronomy 13, 7–11

This theory explains almost everything about religion—except the religious part. It is clear that rituals and sacrifices can bring people together, and it may well be that a group that does such things has an advantage over one that does not. But it is not clear why a *religion* has to be involved. Why are gods, souls, an afterlife, miracles, divine creation of the universe, and so on brought in? The theory doesn't explain what we are most interested in, which is belief in the supernatural.

III. Bodies and Souls

ENTHUSIASM IS BUILDING AMONG SCIENTISTS for a quite different view—that religion emerged not to serve a purpose but by accident.

This is not a value judgment. Many of the good things in life are, from an evolutionary perspective, accidents. People sometimes give money, time, and even blood to help unknown strangers in faraway countries whom they will never see. From the perspective of one's genes this is disastrous—the suicidal squandering of resources for no benefit. But its origin is not magical; long-distance altruism is most likely a by-product of other, more adaptive traits, such as empathy and abstract reasoning. Similarly, there is no reproductive advantage to the pleasure we get from paintings or movies. It just so happens that our eyes and brains, which evolved to react to three-dimensional objects in the real world, can respond to two-dimensional projections on a canvas or a screen.

Supernatural beliefs might be explained in a similar way. This is the religion-as-accident theory that emerges from my work and the work of cognitive scientists such as Scott Atran, Pascal Boyer, Justin Barrett, and Deborah Kelemen. One version of this theory begins with the notion that a distinction between the physical and the psychological is fundamental to human thought. Purely physical things, such as rocks and trees, are subject to the pitiless laws of Newton. Throw a rock, and it will fly through space on a certain path; if you put a branch on the ground, it will not disappear, scamper away, or fly into space. Psychological things, such as people, possess minds, intentions, beliefs, goals, and desires. They move unexpectedly, according to volition and whim; they can chase or run away. There is a moral difference as well: a rock cannot be evil or kind; a person can.

Where does the distinction between the physical and the psychological come from? Is it something we learn through experience, or is it somehow pre-wired into our brains? One way to find out is to study babies. It is notoriously difficult to know what babies are thinking, given that they can't speak and have little control over their bodies. (They are harder to test than rats or pigeons because they cannot run mazes or peck levers.) But recently investigators have used the technique of showing them different events and recording how long they look at them, exploiting the fact that babies, like the rest of us, tend to look longer at something they find unusual or bizarre.

This has led to a series of striking discoveries. Six-month-olds understand that physical objects obey gravity. If you put an object on a table and then remove the table, and the object just stays there (held by a hidden wire), babies are surprised; they expect the object to fall. They expect objects to be solid, and contrary to what is still being taught in some psychology classes, they understand that objects persist over time even if hidden. (Show a baby an object and then put it behind a screen. Wait a little while and then remove the screen. If the object is gone, the baby is surprised.) Five-month-olds can even do simple math, appreciating that if first one object and then another is placed behind a screen, when the screen drops there should be two objects, not one or three. Other experiments find the same numerical understanding in nonhuman primates, including macaques and tamarins, and in dogs.

Similarly precocious capacities show up in infants' understanding of the social world. Newborns prefer to look at faces over anything else, and the sounds they most like to hear are human voices—preferably their mothers'. They quickly come to recognize different emotions, such as anger, fear, and happiness, and respond appropriately to them. Before they are a year old they can determine the target of an adult's gaze, and can learn by attending to the emotions of others; if a baby is crawling toward an area that might be dangerous and an adult makes a horrified or disgusted face, the baby usually knows enough to stay away.

A skeptic might argue that these social capacities can be explained as a set of primitive responses, but there is some evidence that they reflect a deeper understanding. For instance, when twelve-month-olds see one object chasing another, they seem to understand that it really is chasing, with the goal of catching; they expect the chaser to continue its pursuit along the most direct path, and are surprised when it does otherwise. In some work I've done with the psychologists Valerie Kuhlmeier, of Queen's University, and Karen Wynn, of Yale, we found that when babies see one character in a movie help an individual and a different character hurt that individual, they later expect the individual to approach the character that helped it and to avoid the one that hurt it.

Understanding of the physical world and understanding of the social world can be seen as akin to two distinct computers in a baby's brain, running separate programs and performing separate tasks. The understandings develop at different rates: the social one emerges somewhat later than the physical one. They evolved at different points in our prehistory; our physical understanding is shared by many species, whereas our social understanding is a relatively recent adaptation, and in some regards might be uniquely human.

That these two systems are distinct is especially apparent in autism, a developmental disorder whose dominant feature is a lack of social understanding. Children with autism typically show impairments in communication (about a third do not speak at all), in imagination (they tend not to engage in imaginative play), and most of all in socialization. They do not seem to enjoy the company of others; they don't hug; they are hard to reach out to. In the most extreme cases children with autism see people as nothing more than objects—objects that move in unpredictable ways and make unexpected noises and are therefore frightening. Their understanding of other minds is impaired, though their understanding of material objects is fully intact.

At this point the religion-as-accident theory says nothing about supernatural beliefs. Babies have two systems that work in a cold-bloodedly rational way to help them anticipate and understand—and, when they get older, to manipulate—physical and social entities. In other words, both these systems are biological adaptations that give human beings a badly needed head start in dealing with objects and people. But these systems go awry in two important ways that are the foundations of religion. First, we perceive the world of objects as essentially separate from the world of minds, making it possible for us to envision soulless bodies and bodiless souls. This helps explain why we believe in gods and an afterlife. Second, as we will see, our system of social understanding overshoots, inferring goals and desires where none exist. This makes us animists and creationists.

IV. Natural-Born Dualists

For those of us who are not autistic, the separateness of these two mechanisms, one for understanding the physical world and one for understanding the social world, gives rise to a duality of experience. We experience the world of material things as separate from the world of goals and desires. The biggest consequence has to do with the way we think of ourselves and others. We are dualists; it seems intuitively obvious that a physical body and a conscious entity—a mind or soul—are genuinely distinct. We don't feel that we *are* our bodies. Rather, we feel that we *occupy* them, we *possess* them, we *own* them.

This duality is immediately apparent in our imaginative life. Because we see people as separate from their bodies, we easily understand situations in which people's bodies are radically changed while their personhood stays intact. Kafka envisioned a man transformed into a gigantic insect; Homer described the plight of men transformed into pigs; in *Shrek 2* an ogre is transformed into a human being, and a donkey into a steed; in *Star Trek* a scheming villain forcibly occupies Captain Kirk's body so as to take command of the *Enterprise;* in *The Tale of the Body Thief,* Anne Rice tells of a vampire and a human being who agree to trade bodies for a day; and in *13 Going on 30* a teenager wakes up as thirty-year-old Jennifer Garner. We don't think of these events as real, of course, but they are fully understandable; it makes intuitive sense to us that people can be separated from their bodies, and similar transformations show up in religions around the world.

This notion of an immaterial soul potentially separable from the body clashes starkly with the scientific view. For psychologists and neuroscientists, the brain is the source of mental life; our consciousness, emotions, and will are the products of neural processes. As the claim is sometimes put, *The mind is what the brain does.* I don't want to overstate the consensus here; there is no accepted theory as to precisely how this happens, and some scholars are skeptical that we will ever develop such a theory. But no scientist takes seriously Cartesian dualism, which posits that thinking need not involve the brain. There is just too much evidence against it.

Still, it *feels* right, even to those who have never had religious training, and even to young children. This became particularly clear to me one night when I was arguing with my six-year-old son, Max. I was telling him that he had to go to bed, and he said, "You can make me go to bed, but you can't make me go to sleep. It's *my* brain!" This piqued my interest, so I began to ask him questions about what the brain does and does not do. His answers showed an interesting split. He insisted that the brain was involved in perception—in seeing, hearing, tasting, and smelling—and he was adamant that it was responsible for thinking. But, he said, the brain was not essential for dreaming, for feeling sad, or for loving his brother. "That's what *I* do," Max said, "though my brain might help me out."

Max is not unusual. Children in our culture are taught that the brain is involved in thinking, but they interpret this in a narrow sense, as referring to conscious problem solving, academic rumination. They do not see the brain as the source of conscious experience; they do not identify it with their selves. They appear to think of it as a cognitive prosthesis—there is Max the person, and then there is his brain, which he uses to solve problems just as he might use a computer. In this commonsense conception the brain is, as Steven Pinker puts it, "a pocket PC for the soul."

If bodies and souls are thought of as separate, there can be bodies without souls. A corpse is seen as a body that used to have a soul. Most things—chairs, cups, trees—never had souls; they never had will or consciousness. At least some nonhuman animals are seen in the same way, as what Descartes described as "beast-machines," or complex automata. Some artificial creatures, such as industrial robots, Haitian zombies, and Jewish golems, are also seen as soulless beings, lacking free will or moral feeling.

Then there are souls without bodies. Most people I know believe in a God who created the universe, performs miracles, and listens to prayers. He is omnipotent and omniscient, possessing infinite kindness, justice, and mercy. But he does not in any literal sense have a body. Some people also believe in lesser noncorporeal beings that can temporarily take physical form or occupy human beings or animals: examples include angels, ghosts, poltergeists, succubi, dybbuks, and the demons that Jesus so frequently expelled from people's bodies.

This belief system opens the possibility that we ourselves can survive the death of our bodies. Most people believe that when the body is destroyed, the soul lives on. It might ascend to heaven, descend to hell, go off into some sort of parallel world, or occupy some other body, human or animal. Indeed, the belief that the world teems with ancestor spirits—the souls of people who have been liberated from their bodies through death—is common across cultures. We can imagine our bodies being destroyed, our brains ceasing to function, our bones turning to dust, but it is harder—some would say impossible—to imagine the end of our very existence. The notion of a soul without a body makes sense to us.

Others have argued that rather than believing in an afterlife because we are dualists, we are dualists because we want to believe in an afterlife. This was Freud's position. He speculated that the "doctrine of the soul" emerged as a solution to the problem of death;

if souls exist, then conscious experience need not come to an end. Or perhaps the motivation for belief in an afterlife is cultural: we believe it because religious authorities tell us that it is so, possibly because it serves the interests of powerful leaders to control the masses through the carrot of heaven and the stick of hell. But there is reason to favor the religion-as-accident theory.

In a significant study the psychologists Jesse Bering, of the University of Arkansas, and David Bjorklund, of Florida Atlantic University, told young children a story about an alligator and a mouse, complete with a series of pictures, that ended in tragedy: "Uh oh! Mr. Alligator sees Brown Mouse and is coming to get him!" [The children were shown a picture of the alligator eating the mouse.] "Well, it looks like Brown Mouse got eaten by Mr. Alligator. Brown Mouse is not alive anymore."

The experimenters asked the children a set of questions about the mouse's biological functioning—such as "Now that the mouse is no longer alive, will he ever need to go to the bathroom? Do his ears still work? Does his brain still work?"—and about the mouse's mental functioning, such as "Now that the mouse is no longer alive, is he still hungry? Is he thinking about the alligator? Does he still want to go home?"

As predicted, when asked about biological properties, the children appreciated the effects of death: no need for bathroom breaks; the ears don't work, and neither does the brain. The mouse's body is gone. But when asked about the psychological properties, more than half the children said that these would continue: the dead mouse can feel hunger, think thoughts, and have desires. The soul survives. And *children believe this more than adults do,* suggesting that although we have to learn which specific afterlife people in our culture believe in (heaven, reincarnation, a spirit world, and so on), the notion that life after death is possible is not learned at all. It is a by-product of how we naturally think about the world.

V. We've Evolved to Be Creationists

THIS IS JUST HALF THE STORY. Our dualism makes it possible for us to think of supernatural entities and events; it is why such things make sense. But there is another factor that makes the perception of them compelling, often irresistible. We have what the anthropologist Pascal Boyer has called a hypertrophy of social cognition. We see purpose, intention, design, even when it is not there.

In 1944 the social psychologists Fritz Heider and Mary-Ann Simmel made a simple movie in which geometric figures—circles, squares, triangles—moved in certain systematic ways, designed to tell a tale. When shown this movie, people instinctively describe the figures as if they were specific types of people (bullies, victims, heroes) with goals and desires, and repeat pretty much the same story that the psychologists intended to tell. Further research has found that bounded figures aren't even necessary—one can get much the same effect in movies where the "characters" are not single objects but moving groups, such as swarms of tiny squares.

Stewart Guthrie, an anthropologist at Fordham University, was the first modern scholar to notice the importance of this tendency as an explanation for religious thought. In his book *Faces in the Clouds*, Guthrie presents anecdotes and experiments showing that people attribute human characteristics to a striking range of real-world entities, including bicycles, bottles, clouds, fire, leaves, rain, volcanoes, and wind. We are hypersensitive to signs of agency—so much so that we see intention where only artifice or accident exists. As Guthrie puts it, the clothes have no emperor.

Our quickness to over-read purpose into things extends to the perception of intentional design. People have a terrible eye for randomness. If you show them a string of heads and tails that was produced by a random-number generator, they tend to think it is rigged—it looks orderly to them, too orderly. After 9/11 people claimed to see Satan in the billowing smoke from the World Trade Center. Before that some people were stirred by the Nun Bun, a baked good that bore an eerie resemblance to Mother Teresa. In November of 2004 someone posted on eBay a ten-year-old grilled cheese sandwich that looked remarkably like the Virgin Mary; it sold for twenty-eight thousand dollars. (In response pranksters posted a grilled cheese sandwich bearing images of the Olsen twins, Mary-Kate and Ashley.) There are those who listen to the static from radios and other electronic devices and hear messages from dead people—a phenomenon presented with great seriousness in the Michael Keaton movie *White Noise.* Older readers who lived their formative years before CDs and Mpegs might remember listening intently for the significant and sometimes scatological messages that were said to come from records played backward.

Sometimes there really are signs of nonrandom and functional design. We are not being unreasonable when we observe that the eye seems to be crafted for seeing, or that the leaf insect seems colored with the goal of looking very much like a leaf. The evolutionary biologist Richard Dawkins begins *The Blind Watchmaker* by conceding this point: "Biology is the study of complicated things that give the appearance of having been designed for a purpose." Dawkins goes on to suggest that anyone before Darwin who did not believe in God was simply not paying attention.

Darwin changed everything. His great insight was that one could explain complex and adaptive design without positing a divine designer. Natural selection can be simulated on a computer; in fact, genetic algorithms, which mimic natural selection, are used to solve otherwise intractable computational problems. And we can see natural selection at work in case studies across the world, from the evolution of beak size in Galápagos finches to the arms race we engage in with many viruses, which have an unfortunate capacity to respond adaptively to vaccines.

Richard Dawkins may well be right when he describes the theory of natural selection as one of our species' finest accomplishments; it is an intellectually satisfying and empirically supported account of our own existence. But almost nobody believes it. One poll found that more than a third of college undergraduates believe that the Garden of Eden was where the first human beings appeared. And even among those who claim to endorse Darwinian evolution, many distort it in one way or another, often seeing it as a mysterious internal force driving species toward perfection. (Dawkins writes that it appears almost as if "the human brain is specifically designed to misunderstand Darwinism.") And if you are tempted to see this as a red state–blue state issue, think again: although it's true that more Bush voters than Kerry voters are creationists, just about half of Kerry voters believe that God created human beings in their present form, and most of the rest believe that although we evolved from less-advanced life forms, God guided the process. Most Kerry voters want evolution to be taught either alongside creationism or not at all.

What's the problem with Darwin? His theory of evolution does clash with the religious beliefs that some people already hold. For Jews and Christians, God willed the world into being in six days, calling different things into existence. Other religions posit more physical processes on the part of the creator or creators, such as vomiting, procreation, masturbation, or the molding of clay. Not much room here for random variation and differential reproductive success.

But the real problem with natural selection is that it makes no intuitive sense. It is like quantum physics; we may intellectually grasp it, but it will never feel right to us. When we see a complex structure, we see it as the product of beliefs and goals and desires. Our social mode of understanding leaves it difficult for us to make sense of it any other way. Our gut feeling is that design requires a designer—a fact that is understandably exploited by those who argue against Darwin.

It's not surprising, then, that nascent creationist views are found in young children. Four-year-olds insist that everything has a purpose, including lions ("to go in the zoo") and clouds ("for raining"). When asked to explain why a bunch of rocks are pointy, adults prefer a physical explanation, while children choose a functional one, such as "so that animals could scratch on them when they get itchy." And when asked about the origin of animals and people, children tend to prefer explanations that involve an intentional creator, even if the adults raising them do not. Creationism—and belief in God—is bred in the bone.

VI. Religion and Science Will Always Clash

Some might argue that the preceding analysis of religion, based as it is on supernatural beliefs, does not apply to certain non-Western faiths. In his recent book, *The End of Faith,* the neuroscientist Sam Harris mounts a fierce attack on religion, much of it directed at Christianity and Islam, which he criticizes for what he sees as ridiculous factual claims and grotesque moral views. But then he turns to Buddhism, and his tone shifts to admiration—it is "the most complete methodology we have for discovering the intrinsic freedom of consciousness, unencumbered by any dogma." Surely this religion, if one wants to call it a religion, is not rooted in the dualist and creationist views that emerge in our childhood.

Fair enough. But while it may be true that "theologically correct" Buddhism explicitly rejects the notions of body-soul duality and immaterial entities with special powers, actual Buddhists believe in such things. (Harris himself recognizes this; at one point he complains about the millions of Buddhists who treat the Buddha as a Christ figure.) For that matter, although many Christian theologians are willing to endorse evolutionary biology—and it was legitimately front-page news when Pope John Paul II conceded that Darwin's theory of evolution might be correct—this should not distract us from the fact that many Christians think evolution is nonsense.

Or consider the notion that the soul escapes the body at death. There is little hint of such an idea in the Old Testament, although it enters into Judaism later on. The New Testament is notoriously unclear about the afterlife, and some Christian theologians have argued, on the basis of sources such as Paul's letters to the Corinthians, that the idea of a soul's rising to heaven conflicts with biblical authority. In 1999 the pope himself cautioned people to think of heaven not as an actual place but, rather, as a form of existence—that of being in relation to God.

Despite all this, most Jews and Christians, as noted, believe in an afterlife—in fact, even people who claim to have no religion at all tend to believe in one. Our afterlife beliefs are clearly expressed in popular books such as *The Five People You Meet in Heaven* and *A Travel Guide to Heaven.* As the *Guide* puts it,

> Heaven is dynamic. It's bursting with excitement and action. It's the ultimate playground, created purely for our enjoyment, by someone who knows what enjoyment means, because He invented it. It's Disney World, Hawaii, Paris, Rome, and New York all rolled up into one. And it's *forever*! Heaven truly is the vacation that never ends.

(This sounds a bit like hell to me, but it is apparently to some people's taste.)

Religious authorities and scholars are often motivated to explore and reach out to science, as when the pope embraced evolution and the Dalai Lama became involved with neuroscience. They do this in part to make their worldview more palatable to others, and in part because they are legitimately concerned about any clash with scientific findings. No honest person wants to be in the position of defending a view that makes manifestly false claims, so religious authorities and scholars often make serious efforts toward reconciliation—for instance, trying to interpret the Bible in a way that is consistent with what we know about the age of the earth.

If people got their religious ideas from ecclesiastical authorities, these efforts might lead religion away from the supernatural. Scientific views would spread through religious communities. Supernatural beliefs would gradually disappear as the theologically correct version of a religion gradually became consistent with the secular worldview. As Stephen Jay Gould hoped, religion would stop stepping on science's toes.

But this scenario assumes the wrong account of where supernatural ideas come from. Religious teachings certainly shape many of the specific beliefs we hold; nobody is born with the idea that the birthplace of humanity was the Garden of Eden, or that the soul enters the body at the moment of conception, or that martyrs will be rewarded with sexual access to scores of virgins. These ideas are learned. But the universal themes of religion are not learned. They emerge as accidental by-products of our mental systems. They are part of human nature.

ROBERT R. PROVINE

Yawning

FROM THE *AMERICAN SCIENTIST*

Warning: This is an article that will make you yawn. Not because it's stupefying—far from it. But because yawning is so contagious that even reading about it can induce an involuntary yawn. As psychology professor Robert R. Provine shows, that's just one of the fascinating characteristics of this primitive reflex.

IMAGINE A YAWN. You stretch your jaws open in a wide gape, take a deep inward breath, followed by a shorter exhalation, and end by closing your jaws. Ahhh. You have just joined vertebrates everywhere in one of the animal kingdom's most ancient rites. Mammals and most other animals with backbones yawn; fish, turtles, crocodiles, and birds do it. People start yawning very early, offering further evidence of its ancient origins. Yawning is present by the end of the first trimester of prenatal human development and is obvious in newborns.

Yawning is a wonderfully rich topic for anyone interested in the neural mechanisms of behavior. The simple, stereotyped nature of the yawn permits rigorous description, a first step in discovering neural mechanisms. And this application of the "simple systems" approach involves human beings going about their normal activities; there's no need to use bacteria, fruit flies, or nematodes. You can forget about cleaning messy animal cages—much can be learned by experimenting on oneself and observing fellow *Homo sapiens*.

By now, you may be experiencing one of the most remarkable properties of yawning: its contagion. Yawns are so infectious that simply reading or thinking about them can be the vector of an infectious response. The property of contagiousness offers an opportunity to explore the neurological roots of social behavior, face detection, empathy, imitation, and the possible pathology of these processes in autism, schizophrenia, and brain damage.

Fortunately for aspiring students of yawning, the scientific frontiers are near and relatively unpopulated, the result of our tendency to undervalue and neglect the commonplace. Serious science may require no more than a stopwatch, note pad, and pencil. The accessibility of yawning as a problem makes it ideal for what I call "sidewalk neuroscience," a low-tech approach to the brain and behavior based on everyday experience. Whether you follow in my scientific footsteps or simply read along, don't be put off by the primitive tools, simple methods, and behavioral focus. It's easy to be seduced by the trappings of big science and to neglect the extraordinary in our midst.

When I began to study yawning in the 1980s, it was difficult to convince some of my research students of the merits of "yawning science." Although it may appear quirky, my decision to study yawning was a logical extension to human beings of my research in developmental neuroscience, reported in such papers as "Wing-flapping during Development and Evolution." As a neurobehavioral problem, there is not much difference between the wing-flapping of birds and the face- and body-flapping of human yawners.

The Act of Yawning

THE VERB "TO YAWN" is derived from the Old English *ganien* or *ginian*, meaning to gape or open wide (chasms really *do* yawn). But in addition to gaping jaws, yawning has significant features that are easy to observe and analyze. I collected yawns to study by tapping the contagion response. Back in the 1980s I asked subjects sitting in an isolation chamber to "think about yawning," and to push a button at the start of a yawn and keep it depressed until they finished exhaling at the end of a yawn. (Self-report was used because yawning is inhibited in subjects who think they are being observed.)

Here are some of the things I learned: The yawn is highly stereotyped but not invariant in duration and form. It is an excellent example of the instinctive "fixed action pattern" of classical animal-behavior study, or ethology. It is *not* a reflex, a short-duration, rapid, proportional response to a simple stimulus. But once started, a yawn progresses with the inevitability of a sneeze. The yawn runs its course over about six seconds on average, but its duration can range from about three and a half seconds to much longer than the average. There are no half-yawns, an example of the "typical intensity" of fixed action patterns and a reason why you cannot stifle a yawn. Yawns come in bouts, with a highly variable inter-yawn interval of around sixty-eight seconds. There is no relation between yawn frequency and duration; producers of short or long yawns do not compensate by yawning more or less often.

I offer three informative yawn variants that test hypotheses about the form and function of yawning. If you are now yawning, you can test yourself and draw your own conclusions about yawning and its underlying mechanism. Not everyone, including my long-suffering wife, shares my enthusiasm for such self-experimentation. And even enthusiasts may want to conduct these experiments in private. Let's begin.

The closed-nose yawn. When you feel yourself start to yawn, pinch your nose closed. Most subjects report being able to perform perfectly normal closed-nose yawns. This indicates that the inhalation at the onset of a yawn, and the exhalation at its terminus, need not involve the nostrils—the mouth provides a sufficient airway.

Now let's test some propositions about the role of the mouth and jaw.

The clenched-teeth yawn. When you feel yourself begin to yawn, clench your teeth but permit yourself to inhale normally through your open lips and clenched teeth. This diabolical variant gives one the sensation of being stuck in mid-yawn. It shows that the gaping of the jaws is an essential component of the complex motor program of the yawn; unless it is accomplished, the program will not run to completion. The yawn is also shown to be more than a deep breath because, unlike normal breathing, inhalation and exhalation cannot be performed as well through the clenched teeth as through the nose.

The nose yawn. This variant tests the adequacy of the nasal airway to sustain a yawn. (The closed-nose yawn already showed the nasal airway to be unnecessary for yawns.) Unlike normal breathing, which can be performed equally well through mouth or nose, yawning is impossible via nasal inhalation alone. As with the clenched-teeth yawn, the nose-yawn provides the unfulfilling sensation of being stuck in mid-yawn. Inhalation through the mouth is an essential component of the motor pattern of yawning. Exhalation, on the other hand, can be accomplished equally well through nose or mouth.

So far, you and I have demonstrated that inhalation through the oral airway and the gaping of the jaws are essential for normal yawns, and that the motor program for yawning will not run to completion without feedback that these parts of the program have been accomplished. But yawning is a powerful, generalized movement that involves much more than airway maneuvers and jaw gaping. When yawning, you also stretch your facial muscles, tilt your head back, narrow or close your eyes, tear, salivate, open the eustachian tubes of your middle ear and perform many other, yet unspecified, cardiovascular, neuromuscular, and respiratory acts. Perhaps yawning shares components with other behavior, all being assembled from a neurological parts bin of ancient motor programs. For example, is the yawn a kind of "slow sneeze," or is the sneeze a "fast yawn"? Both share common respiratory and motor features, including jaw gaping, eye closing, and head tilting.

Looking at other kinds of behavior that use some of the same components, we might ask: Does the yawnlike facial expression during sexual climax suggest that the two acts share a neurobehavioral heritage? This proposition is not as far-fetched as it sounds on first hearing, because yawning is triggered by androgens and oxytocin and is associated with other sex-related agents and acts. Doctoral student Wolter Seuntjens at the Vrije Universiteit Amsterdam tracked down these connections when he surveyed the surprisingly extensive but scattered literature on this subject for his dissertation in art history, published last year. Among most mammals, males are the leading yawners. Our species is unique in that both sexes yawn equally often and are sexually receptive at all times. In rats, most chemical agents that produce yawning and stretching also produce penile erection. Although such antidepressant drugs as clomipramine (Anafranil) and fluoxetine (Prozac) typically depress sexual desire and performance, in a few people they have the interesting side effect of producing yawns that trigger orgasms.

Most human yawners are not rewarded with orgasms. But yawning does feel good to most people, being rated 8.5 on a 10-point hedonic scale (1=bad, 10=good). Given the similarities among sexual orgasm, yawning, and sneezing (including some resemblances among the typical facial expressions), it's perfectly reasonable to refer to the resolution of all three acts as a "climax." Is the frustration of being unable to resolve building sexual tension in climax akin to the unsatisfying sensation of being stuck in mid-yawn or mid-sneeze? The chronic urge but inability to yawn is quite disturbing to those who experience it; several people with this problem have contacted me about gaining relief. Given these insights, you can now view your yawning and sneezing friends in a completely different light.

Yawning and stretching also share properties and may be performed together as parts of a global motor complex. But they do not always co-occur—people usually yawn when we stretch, but we don't always stretch when we yawn, especially before bedtime. Studies by J. I. P. deVries, G. H. A. Visser, and H. F. Prechtl in the early 1980s, charting movement in the developing fetus using ultrasound, observed not just yawning but a link between yawning and stretching as early as the end of the first prenatal trimester.

The most extraordinary demonstration of the yawn-stretch

linkage occurs in many people paralyzed on one side of their body because of brain damage caused by a stroke. The prominent British neurologist Sir Francis Walshe noted in 1923 that when these hemiplegics yawn, they are startled and mystified to observe that their otherwise paralyzed arm rises and flexes automatically in what neurologists term an "associated response." Yawning apparently activates undamaged, unconsciously controlled connections between the brain and the cord motor system innervating the paralyzed limb. It is not known whether the associated response is a positive prognosis for recovery, nor whether yawning is therapeutic for reinnervation or prevention of muscular atrophy.

Clinical neurology offers other surprises. Some patients with "locked-in" syndrome, who are almost totally deprived of the ability to move voluntarily, can yawn normally. The neural circuits for spontaneous yawning must exist in the brain stem near other respiratory and vasomotor centers, because yawning is performed by anencephalics who possess only the medulla oblongata. The multiplicity of stimuli of contagious yawning, by contrast, implicates many higher brain regions.

The Folklore of Yawning

Having considered the motor act of yawning, I'm ready to test some of the folklore about when and why we yawn. Although often in error, folklore poses interesting questions and is the repository of centuries of informal observations about human nature. A disadvantage of testing folklore is that when it's confirmed, you're accused of proving the obvious. Research has sometimes confirmed and extended common beliefs about yawning, but my colleagues and I have been rewarded with plenty of surprises.

We yawn when bored. Score a point for folklore. Bored people really do yawn a lot. To induce ennui, I asked subjects to watch a television test pattern for thirty minutes while giving those in the control condition a thirty-minute dose of music videos. However you feel about music videos, you will find them more interesting (less boring) than an unchanging color-bar test pattern. Subjects yawned about 70 percent more during the test pattern than during the music video condition. But yawning is not exclusive to the bored; there is anecdotal evidence of yawning by paratroopers before their first parachute jump, Olympic athletes before their event, a violinist waiting to go on stage to perform a concerto, and dogs on the threshold of attack.

We yawn when sleepy. As expected, subjects who recorded their yawning and sleeping in a diary during a one-week period confirmed that people really do yawn most when sleepy, especially during the hour after waking, and second most during the pre-bedtime hour. A surprise came from accompanying data about stretching that subjects also recorded in their diaries. After waking, subjects simultaneously yawned and stretched. But before bedtime, most subjects only yawned. You can observe this yawn-stretch linkage in your family dog or cat as it rouses from its slumber.

We yawn because of a high level of carbon dioxide or a shortage of oxygen in the blood or brain. These legendary but unsupported factoids are repeated so often that they have a life of their own, still being presented in the popular media and in medical-school lectures. Yet the only test of these hypotheses, one that I conducted eighteen years ago, rejected them. Breathing levels of carbon dioxide a hundred or more times greater than the concentration in air (3 percent or 5 percent CO_2 versus the usual 0.03 percent CO_2) did not increase yawning, although subjects did dramatically increase their breathing rate and tidal volume. Furthermore, breathing 100 percent oxygen did not inhibit yawning.

Although breathing and yawning both involve respiratory acts and are produced by neurological motor programs, these programs are separate and can be modulated independently. An exercise task, for example, that had subjects huffing and puffing at high rates did not affect their rate of ongoing yawning. Test this proposition during your next jog (but not sprint). Thoughts about yawning will prompt yawns that can be sustained independent of how labored your breathing becomes.

The Contagiousness of Yawning

The legendary infectiousness of yawning can be confirmed with simple observation, but the full story about contagion is so broad and deep that it deserves extended treatment. Before exploring its exciting implications for social biology, I will quantify the contagiousness of yawning and define its stimulus triggers.

In early experiments, I tested the contagiousness of yawning by exposing subjects to a five-minute series of thirty videotaped repetitions of a male adult yawning. Subjects were more than twice as likely to yawn while observing yawns (55 percent) as to yawn while viewing a comparable series of smiles (21 percent). Unlike the response of a reflex to a stimulus, the visual stimulus was not followed by a short and predictable latency period; instead, yawning took place throughout the five-minute test period. In the language of classical ethology, the yawning face is a sign stimulus, which activates an innate mechanism that releases the fixed action pattern of the yawn.

The yawn video proved equally potent whether viewed right side up, sideways, or upside down. Furthermore, the subjects' yawn detector was neither color nor movement dependent, because the video was equally potent when viewed in black and white or in color, or when the usually animate stimulus was presented as a still image of the yawner in mid-yawn.

Next I tested what features of the yawning face were most potent in prompting yawns. I even hoped to be able to engineer a supranormal stimulus—the mother of all yawn stimuli. Here the plot thickened. Most people incorrectly presume that the gaping mouth is the essential signature of the yawn. Yet it turned out that yawning faces that had been edited to mask the mouth were just as effective in producing yawns as the intact face. I was skeptical of this finding until complementary data showed that the disembodied yawning mouth was no more effective in evoking yawns than the control smile.

Outside the context of the yawning face, it seems that the gaping mouth is an ambiguous stimulus—the mouth could equally well be yelling or singing. The detector seems to be responding to the overall pattern of the yawning face and upper body, not a particular facial feature. Miss Manners, take note: This incidental contribution to etiquette research suggests that shielding your mouth is a polite but futile gesture that will not prevent the passing of your yawn to others.

Nature conspires to spread yawns. Even thinking about yawning, the yawn-induction procedure used in several of my studies, evoked yawns in 88 percent of subjects within thirty minutes. And as many readers have noticed by now, reading about yawning triggers yawns. When put to the test, 30 percent of subjects who read an article about yawning for five minutes reported yawning during this period, versus 11 percent of a control group who read an article about hiccuping. When the criterion was relaxed to include those who either yawned or were tempted to yawn, the difference between the yawn and hiccup conditions grew to 75 percent and 11 percent, respectively.

My plan to develop an ultra-potent yawn stimulus was abandoned when I discovered the global nature of yawn triggers. I realized that synthesizing a gaping mouth of just the right size that opened and closed at just the right rate would not produce the perfect, irresistible stimulus. The observation of a naturally yawning person does just fine by itself. Also, otherwise neutral stimuli can acquire yawn-inducing properties through association. My reputation as a yawn sleuth has conferred a curious kind of charisma—I've become a yawn stimulus.

The Roots of Sociality

YAWNS ARE REPLICATED BY OBSERVERS who pass their yawns along from one to another in a behavioral chain reaction. This synchronizes the behavioral and physiological state of a group. The underlying mechanism of the contagious response probably involves some sort of neurological detector for yawns, but the wide range of yawn-producing stimuli suggests that the detector must be broadly tuned. Contagious yawning definitely does not involve a conscious desire to replicate the observed act ("I think I'll yawn just like that person did")—we yawn whether we want to or not. Contagious yawning is species-typical, neurologically programmed social behavior of a sort neglected by social scientists, who usually emphasize the role of the environment in shaping the behavior of individuals. The involvement of an unconsciously controlled human universal broadens the discussion of a variety of social behavior.

Consider the presumed imitation of facial expressions by human neonates, first reported by Andrew Meltzoff and M. Keith Moore in a famous 1977 article. This phenomenon continues to be reported in terms of higher-order cognitive processes; we believe that newborn babies are doing some impressive processing because they imitate the faces they see. But is this imitation better evidence of cognition than contagious yawning, which does not involve the intent to model? Contagious laughter, the basis of the notorious television laugh tracks, is another compelling case of unconsciously controlled pseudo-imitative behavior.

The path of inquiry becomes a bit more challenging when one moves from the facts of behavior to speculating about underlying mechanisms. Renewed interest in contagious behavior has been prompted by the discovery of so-called mirror neurons, which have been implicated in a variety of imitative activities ranging from motor control to empathy. Mirror neurons are active both when an act such as grasping is performed and when that same act is observed in others.

But mirror neurons may fall short as a mechanism of contagious yawning because their activity does not trigger an imitative motor act. Recently Steven Platek of Drexel University and colleagues Feroze B. Mohamed of Temple University and Gordon G. Gallup Jr. of the State University of New York at Albany used functional magnetic-resonance imaging to study activity in people exposed to yawns. They found unique activity in the posterior cingulate and precuneus brain regions, areas *not* associated with mirror activity. These regions are, instead, associated with such self-processing functions as self-reference, theory of mind and autobiographical memory. In an unconscious way, someone who "catches" a yawn may be expressing a primal form of empathy.

The sociality inherent in contagious yawning thus may provide a novel marker and diagnostic tool for empathetic responses during evolution and development and in pathology. Little is known about contagiousness beyond the human species. However, last year James Anderson of the University of Stirling, Masako Myowa-Yamakoshi of the University of Shiga Prefecture, and Tetsuro Matsuzawa of Kyoto University demonstrated contagious yawning in adult chimpanzees, a primate that shows rudimentary empathy and self-awareness (as reflected in mirror-recognition tasks). Contagion, if present, may be weaker in monkeys and other animals that are relatively deficient in these traits.

Although spontaneous yawning is performed by human fetuses in the womb, James Anderson and his colleague Pauline Meno at Stirling did not detect contagious yawning in children until several years after birth. This, combined with its rarity in other species, suggests that contagious yawning has a separate and relatively recent evolutionary origin.

In certain neurological and psychiatric disorders, including schizophrenia and autism, the patient has an impaired ability to infer the mental states of others. The evaluation of contagious yawning in such disorders offers substantial rewards. People who are not clinically ill but are schizotypal—that is, who are deficient in their ability to infer or empathize with what others want, know or intend to do and have certain other problems in thought and behavior—were found by Platek and his colleagues to have reduced susceptibility to contagious yawning. Provocatively, the late Canadian psychiatrist Heinz Lehmann claimed that increases in yawning (contagious yawning was not specifically examined) could predict recovery in schizophrenia. Finally, vigilant comatose patients offer an intriguing test of the contagious yawn as a measure of sociality. While these comatose patients yawn, it is unknown whether they can yawn contagiously, or the extent to which this ability correlates with their neurological status and prognosis.

Judgment about contagiousness as evidence for a trait of sociality must, then, await further research. The contagiousness of species-typical acts such as yawning and vocalizations such as laughter and crying may either tap a primitive neurological substrate for social behavior or be confined to that specific behavior and not reflect a more general social process.

Unconscious Control

YOU CANNOT YAWN on command. This observation is the best evidence of yawning's unconscious control. Yawns occur either spontaneously or as a contagious response to an observed yawn. Intense self-awareness, as when you are being observed or even suspect that you may be observed, inhibits yawning.

I already had experimental evidence of this inhibition when I began studying yawning contagion, providing the rationale for the use of self-report in experiments. As my yawning studies attracted attention (the popular media have a voracious appetite for stories on this topic), I had the experience of seeing the inhibition play out.

A television news-magazine crew turned up one day to tape a segment. Against my advice, the show's producer set out to re-create my experiment in which one-half of a large lecture class read an article about yawning while the other half read a control passage about hiccuping. Normally the effect of the yawning article is robust, and it has been used as a demonstration of contagion in classes at other universities.

As I predicted, the demonstration did not survive up-close-and-personal scrutiny by a national network television crew. With the cameras rolling as the students read, only a tiny fraction of the usual amount of yawning was observed. The video crew performed an unintentional but informative variant of my original research demonstrating the powerful effect of social inhibition on yawning. Even highly motivated and prolific yawners who volunteered to be on national television stopped yawning when placed before the camera. It is notable that the social inhibition of yawning occurred unconsciously and was not the voluntary effort of the yawner to suppress a rude or inappropriate act. A socially significant act can be either produced or inhibited by unconscious processes.

Scrutiny also inhibits hiccuping, an act that is also unconscious but is not contagious. When piano students in my wife's home studio start hiccuping, she signals me to bring my tape recorder into the studio to record their sounds. In all nine cases where we have used this technique, my appearance with recorder and microphone has been followed almost immediately by an end to the hiccuping. I've thus discovered an effective treatment for the hiccups while finding further evidence for the social inhibition of an unconsciously controlled act. When the ancient and the new, the unconscious and the conscious compete for the brain's channel of expression, the more modern, conscious mechanism dominates, suppressing its older, unconscious rival.

Conclusion

Too little is known for this article to end with a dazzling intellectual flourish and a Grand Unified Theory of Yawning. It is customary at this point, though, to suggest a need for further study, and indeed I see much potential in using yawning to develop and test theories of mind and to better understand certain neuro- and psychopathologies.

Here I have attempted to describe the yawn, when we do it and its promise for study, without speculating about its function. Yawning appeared very early in vertebrate history, with contagiousness evolving much later. Yawning has many consequences, including opening of the eustachian tube, tearing, inflating the lungs, stretching and signaling drowsiness, but these may be incidental to its primal function—which may be something as unanticipated as sculpting the articulation of the gaping jaw during embryonic development.

Selecting a single function from the many options may be an unrealistic goal. However, reviewing the disparate facts, I'm impressed that yawning is associated with the change of behavioral state—wakefulness to sleep, sleep to wakefulness, alertness to boredom, threshold of attack, sexual arousal, switching from one kind of activity to another. Yawning is a vigorous, widespread act that may stir up our physiology and facilitate these transitions, with the motor act becoming the stimulus for the more recently evolved contagious response.

Consider the Bakairi people of central Brazil as observed by their first European visitor, nineteenth-century ethnologist Karl von den Steinen. Irenäus Eibl-Eibesfeldt recalls in his 1975 book *Ethology* that Steinen reported: "If they seemed to have had enough of all the talk, they began to yawn unabashedly and without placing their hands before their mouths. That the pleasant reflex was contagious could not be denied. One after the other got up and left until I remained with my dujour." Among all members of our species, the chain reaction of contagious yawning synchronizes the behavioral as well as the physiological state of our tribe. Yawning is a reminder that ancient and unconscious behavior lurks beneath the veneer of culture, rationality and language, continuing to influence our lives.

BIBLIOGRAPHY

Anderson, J. R., and P. Meno. 2003. Psychological influences on yawning in children. *Current Psychology Letters* 11. http://cpl.revues.org/document390.html.

Anderson, J. R., M. Myowa-Yamakoshi, and T. Matsuzawa. 2004. *Proceedings of the Royal Society of London B* 271(Suppl. 6): S468–S470.

Baenninger, R. 1997. On yawning and its functions. *Psychonomic Bulletin and Review* 4:198–207.

DeVries, J. I., G. H. A. Visser, and H. F. R. Prechtl. 1982. The emergence of fetal behavior: I. Qualitative aspects. *Early Human Development* 7:301–322.

Eibl-Eibesfeldt, I. 1975. *Ethology* (2nd ed.). New York: Holt, Rinehart and Winston.

Meltzoff, A. N., and M. K. Moore. 1977. Imitation of facial and manual gestures by human neonates. *Science* 198:75–78.

Platek, S. M., F. B. Mohamed, and G. G. Gallup. 2005. Contagious yawning and the brain. *Cognitive Brain Research* 23:448–452.

Provine, R. R. 1986. Yawning as a stereotyped action pattern and releasing stimulus. *Ethology* 72:109–122.

Provine, R. R. 1989. Faces as releasers of contagious yawning: An approach to face detection using normal human subjects. *Bulletin of the Psychonomic Society* 27:211–214.

Provine, R. R. 2000. *Laughter: A Scientific Investigation.* New York: Viking.

Provine, R. R. 2005. Contagious yawning and laughing: Everyday imitation- and mirror-like behavior (p. 146). Commentary on: Arbib, M. A. 2005. From monkey-like action recognition to human language: An evolutionary framework for neurolinguistics. *Behavioral and Brain Sciences* 28:105–167.

Provine, R. R., and H. B. Hamernik. 1986. Yawning: Effects of stimulus interest. *Bulletin of the Psychonomic Society* 24:437–438.

Provine, R. R., B. C. Tate, and I. Geldmacher. 1987. Yawning: No effect of 3–5% CO_2, 100% O_2, and exercise. *Behavioral and Neural Biology* 48:382–93.

Provine, R. R., H. B. Hamernik, and B. C. Curchack. 1987. Yawning: Relation to sleeping and stretching in humans. *Ethology* 76:152–160.

Seuntjens, W. 2004. On yawning or the hidden sexuality of the human yawn. Dissertation. Vrije Universiteit Amsterdam.

Walshe, F. M. R. 1923. On certain tonic or postural reflexes in hemiplegia with special reference to the so-called "associative movements." *Brain* 46:39–43.

Walusinski, O. 2005. Le bâillement. Online dossier.http://www.baillement.com.

KENNETH CHANG

Ten Planets? Why Not Eleven?

FROM THE *NEW YORK TIMES*

While the Michael Brown who dominated the headlines in 2005 achieved notoriety, it may be another Michael Brown—a scientist— who will be long remembered after Hurricane Katrina is forgotten. Kenneth Chang meets the planetary astronomer who has discovered what may well be our solar system's tenth planet.

Between feedings and diaper changes of his newborn daughter, Michael E. Brown may yet find an eleventh planet.

Once conducted almost exclusively on cold, lonely nights, observational astronomy these days is often done under bright California sunshine.

When he has a few spare minutes, Dr. Brown, a professor of planetary astronomy at the California Institute of Technology, downloads images taken during a previous night by a robotically driven telescope at Palomar Observatory one hundred miles away. Each night, the telescope scans a different swath of sky, photographing each patch three times, spaced an hour and a half apart.

In any one of the photographs, a planet or some other icy body at the edge of the solar system looks just like a star. Unlike a star it moves between the exposures.

Dr. Brown's computer programs flag potential discovery candidates for him to inspect. He quickly dismisses almost all of them—double images caused by a bumping of the telescope, blurriness from whirls in the atmosphere or random noise.

Sometimes, like last January 5, he spots a moving dot.

Dr. Brown had rewritten his software to look for slower-moving and more distant objects.

On that morning, he was sitting in his Caltech office—unremarkable university turf sparsely decorated with a not-full bottle of Jack Daniel's, a dragon mobile, a dinosaur toothbrush, a Mr. Potato Head, and other toys and knickknacks that long predated parenthood—and reexamining images from nearly a year and a half earlier, October 21, 2003.

The first several candidates offered by the computer were the usual garbled images.

Then he saw it: a bright, unmistakably round dot moving across the star field.

He did a quick calculation. Even if this new object reflected 100 percent of the sunlight that hit it—and nothing is perfectly reflective—it would still be almost as large as Pluto. That meant, without any additional data, Dr. Brown knew he had discovered what could be the tenth planet.

He noted the time: 11:20 A.M. Dr. Brown knew that astrologers would ask because they had asked after earlier discoveries of smaller Kuiper Belt objects.

He thought about a bet he had made with a friend, for five bottles of Champagne, that he would discover something larger than Pluto by January 1, 2005, five days earlier. He sent her an e-mail message asking for a five-day extension.

Dr. Brown seems destined for future astronomy textbooks, either as the discoverer of the first new planet in seventy-five years or as the man who pushed Pluto out of the planetary pantheon. Astronomers have long argued over whether Pluto should be a planet and speculated that ice balls larger than Pluto might be hiding in the Kuiper Belt, a ring of debris beyond the orbit of Neptune. Several objects approaching the size of Pluto have been discovered in recent years.

Dr. Brown and two colleagues, David Rabinowitz of Yale and Chad Trujillo of the Gemini Observatory in Hawaii, are the first to point to something that is almost certainly larger than Pluto.

"If people want to get rid of Pluto, I'm more than happy to get rid of Pluto and say this one isn't a planet, either," Dr. Brown said.

"If culturally we would be willing to accept a scientific definition, that would be great," he continued. "The only thing that would make me unhappy is if Pluto remained a planet, and this one was not one."

So far, little is known about the new planet, which carries the temporary designation of 2003 UB313. It is currently 9 billion miles from the sun, at the farthest point of its five-hundred-sixty-year orbit.

A couple of centuries from now, its elliptical trajectory will take it within 3.3 billion miles of the sun, closer than some of Pluto's orbit. And the orbit is surprisingly askew.

While most of the solar system circles the sun in a flat disc, the new planet's orbit is tilted about 45 degrees from the disc.

Planet finding was not a career goal for Dr. Brown. Until recently, he was among those who argued that nothing in the Kuiper Belt deserved to be called a planet, not even Pluto.

Dr. Brown, forty, grew up in Alabama, a child of the 1970s, which left a mark in his speech patterns. For things he likes, he inevitably calls them "cool."

His father, an engineer at IBM, had moved to Huntsville to work on the giant Saturn rockets, which were being designed at NASA's Marshall Space Flight Center to carry astronauts to the moon.

When he was finishing up his undergraduate degree in physics at Princeton, he thought he would pursue theoretical work in cosmology, devising ideas about how the universe came together.

Then James Peebles, a physics professor at Princeton, mentioned to him how astronomy needed more observers actually looking at the sky.

"That was it," Dr. Brown recalled. "As soon as he said it, I was like, okay."

In graduate school at the University of California, Berkeley, he still planned to work on the far, far away—distant galaxies that blast out loud radio signals. His thesis adviser, Hyron Spinrad, made his graduate students also work on comets, because he was interested in them and could not get help otherwise. Dr. Brown was captivated as well. "I thought, This stuff is cool," he said.

That brought his interests inside the solar system. Later Dr. Brown came across a tiny, little-used twenty-four-inch telescope at the Lick Observatory outside San Jose. "You find the telescope first and then find your thesis program," he said.

His thesis topic turned out to be the volcanoes on Io, one of the moons of Jupiter, and how the gases from the volcanoes are swept up into Jupiter's magnetic fields, accelerated into orbit around Jupiter, and then slammed back into Io at 125,000 miles per hour.

He entered the planet-searching business through a chance opportunity. When he arrived at Caltech a decade ago, the forty-eight-inch Samuel Oschin Telescope at Palomar was just finishing up a large sky survey. While many astronomers clamor for use of Palomar's main two-hundred-inch telescope, Dr. Brown realized he could easily get ample time on the smaller one.

So, just as he did at Lick, he looked for a project to fit the telescope. The first Kuiper Belt object had been discovered a few years earlier, and Dr. Brown thought it would be useful to do a systematic sweep of the sky to look for them. In past centuries, the trick to discovering a planet was knowing exactly where to look. Now, computers and automated telescopes have allowed a new strategy: look everywhere.

Dr. Brown has a love of code names, so he and his colleagues decided if they ever found something larger than Pluto, they would give it the code name Xena, after the television series starring Lucy Lawless as an ancient Greek warrior.

It was also partly a nod to Planet X, a long-hypothesized massive planet in the outer solar system.

In 2002 Dr. Brown and Dr. Trujillo turned up their first big find: a Kuiper Belt object about 775 miles wide, or about as large as Pluto's moon, Charon. They named it Quaoar, after a god in Native American mythology, and learned their first lesson in astronomical discovery: find a pronounceable name. (Quaoar is pronounced KWAH-o-wahr.)

A year later, Dr. Brown, Dr. Trujillo, and Dr. Rabinowitz of Yale found something stranger, an object larger than Quaoar and much farther out, 8 billion miles. And stranger yet, its eleven-hundred-year orbit carries it out as far as 84 billion miles, far beyond anything else known in the solar system.

Initially, they excitedly thought it might be larger than Pluto and called it Xena for a few days. But further measurements and calculations indicated it was probably, at most, three-quarters the diameter of Pluto. When they announced the discovery last year, they gave it the name of Sedna, after an Inuit goddess.

The name was easily pronounceable, but it peeved some astronomers, especially amateur asteroid searchers, who felt Dr. Brown had again flouted the International Astronomical Union's naming rules. The rules prohibit discoverers from publicly announcing any name, even a tentative one, before the union approves it, a process that takes months to years. The critics wanted the union to rebuke Dr. Brown and reject the name.

One amateur astronomer, Reiner M. Stoss, even proposed Sedna as a name for a small, earlier discovered asteroid, to thwart Dr. Brown.

Dr. Brown admitted, "I consciously broke the rules," but he felt that the preliminary designation, 2003 VB12, was too obscure and confusing for a noteworthy addition to the solar system. "I thought, this is stupid," he said. "This object needs to have a name before it goes public."

After Sedna, the astronomers also realized that there could be more discoveries lurking in their photographs.

Dr. Brown finds that betting is an effective way to spur scientific progress. He would have found the planet sooner or later, but late last year he realized that his best hope for winning the Champagne bet was to sift through thousands of candidates in the old images.

On December 20, going through the images from the previous May, he discovered a new bright Kuiper Belt object.

It was big, but not big enough. He gave that one the code name of Santa.

When he later discovered a moon around Santa, he gave the moon the code name Rudolph. (When he discovered yet another Kuiper Belt object in April, he continued in the holiday vein and code-named that one Easter Bunny. He even had a code name for his daughter before she was born and he and his wife had not yet settled on a name. The whiteboard in his office had a list of tasks labeled "TDBP"—To Do Before Petunia.)

He spent Christmas through New Year's examining old images.

Nothing turned up.

Then on January 5—not January 8 as he had said at his news conference—he finally found one he could call Xena. "What if it's the size of Mercury?" he mused in his notes.

This time, Dr. Brown decided to play by the International Astronomical Union's rules and did not announce his real intended name for 2003 UB313, leading to rumors that he had officially proposed Xena.

Meanwhile, Dr. Brown, on family leave until the end of the year, found a new set of data to work with: his daughter Lilah's sleeping and eating patterns.

In the hospital, he and his wife had, like many new parents, written down a record of feeding and sleeping times.

Dr. Brown wrote a computer program to generate colorful charts from the information and put them online at lilahbrown.com.

"Who wouldn't be fascinated?" Dr. Brown asked. "Well, I guess, most people."

Black bars indicate the times that his wife, Diane Binney, is breastfeeding Lilah. Blue bars indicate when Dr. Brown is feeding a bottle to her. Green bars indicate the hours Lilah is awake and content; red bars indicate the fussy times.

Dr. Brown said he hoped to find a pattern in Lilah's sleep cycle—

she has been waking up, on average, every 2.5 hours. If one nap lasted only 1.5 hours, would that mean the next nap would last 3.5 hours, giving her parents a chance to rest?

A chart with cloudlike splatter of data points gave the unfortunate answer. “There is absolutely no correlation,” Dr. Brown said.

RICHARD PRESTON

Climbing the Redwoods

FROM *THE NEW YORKER*

The redwood forests of the Pacific Northwest are one of America's natural wonders. Ancient, tall as skyscrapers, redwoods are a marvel. Perhaps most remarkable is the redwood canopy, a living world three hundred feed above ground. Richard Preston joins a preeminent botanist for a first-hand tour of the redwoods' peaks.

THE COAST REDWOOD TREE is an evergreen conifer, a member of the cypress family, which grows in valleys and on slopes of mountains along the coast of Central and Northern California, mostly within ten miles of the sea. The scientific name of the tree, which is usually simply called a redwood, is *Sequoia sempervirens.* A coast redwood has fibrous, furrowed bark, flat needles, and small seed-bearing cones the size of olives. Its heartwood is the color of old claret and is extremely resistant to rot. It has a lemony scent. Redwoods flourish in wet, rainy, foggy habitats. The realm of the redwoods starts in Big Sur and runs northward along the coast to Oregon; fourteen and a half miles up the Oregon coast, the redwoods abruptly stop.

The main trunk of a coast redwood can be up to twenty-five feet in diameter near its base, and in some cases it can extend upward from the ground for more than two hundred and fifty feet before the first strong branches emerge and the crown of the tree begins to flare. The crown of a tall coast redwood is typically an irregular spire that can look like the plume of a rocket taking off. Very few trees of any species today other than redwoods are more than three hundred feet tall. The tallest living coast redwoods are between three hundred and fifty and three hundred and seventy feet high—the equivalent of a thirty-five-to-thirty-seven-story building.

In its first fifty years of life, a coast redwood can grow from a seed into a tree that's a hundred feet tall. Redwoods grow largest and tallest in silty floodplains near creeks, in spots that are called alluvial flats. There, a redwood can suck up huge amounts of water and nutrients, and it is protected from wind, which can throw a redwood down. In its next thousand years, it grows faster, adding mass at an accelerating rate. The living portion of a tree is a layer of tissue called the cambium, which exists underneath the bark. If the cambium of one of the bigger coast redwoods was to be spread out into a flat sheet, it would be nearly half the size of a football field. Each year, a coast redwood can add one millimeter of new growth to its cambium, or the equivalent of one ton of new wood.

As a young redwood reaches maturity, it typically loses its top. The top either breaks off in a storm or dies and falls off. A redwood reacts to the loss by sending out new trunks, which typically appear in the crown, high up in the tree, and point at the sky like the fingers of an upraised hand. The new trunks shoot upward from larger limbs, traveling parallel to the main trunk, or they emerge directly out of the main trunk and run alongside it. The new trunks send out their own branches, which eventually spit out more trunks through the crown. The resulting structure is what mathematicians call a fractal; botanists say that the tree is forming reiterations. The redwood is repeating its shape again and again.

With the passing of centuries, the reiterated trunks begin to touch one another here and there. The trunks fuse and flow together at these spots, like Silly Putty melting into itself. The bases of the extra trunks bloat out, and become gnarled masses of living wood called buttresses. In the crowns of the largest redwoods, bridges of living redwood are flung horizontally from branch to branch and from trunk to trunk, cross-linking the crown with a natural system of struts and cantilevers. This strengthens the crown and may help it to grow bigger, until it can look like a thunderhead coming to a boil. There are often blackened chambers and holes in the trunks—fire caves, caused by big forest fires. The tree survives and regrows its burned parts, and it continues to thrust out new trunks.

Botanists judge the size of a tree by the amount of wood it contains, not by its height. By that measure, the largest species of tree is the giant sequoia, a type of cypress that is closely related to the coast redwood. The biggest living giant sequoia trees have fatter and more massive trunks than the coast redwoods. But the coast redwood is the tallest species of tree on earth.

Extremely large coast redwoods are referred to as redwood giants. The very biggest are called titans. Currently, about a hundred and twenty coast redwoods are known to be more than three hundred and fifty feet tall. Eighty percent of them live in Humboldt Redwoods State Park, along the Eel River, in Humboldt County, in Northern California. No one knows exactly how old the biggest coast redwoods are because nobody has ever drilled into one of them to count its annual growth rings. Botanists think that the oldest redwoods may be somewhere between two thousand and three thousand years old. They seem to be roughly the age of the Parthenon.

ONE DAY IN EARLY 2003 I arrived at the office of Stephen C. Sillett, on the campus of Humboldt State University in Arcata, a coastal town in Humboldt County. Winter storms had been passing through, and Arcata smelled of redwoods and the sea. Sillett is an associate professor of botany, and he is the principal explorer of the redwood-forest canopy, the three-dimensional labyrinth that exists in the air above the forest floor. He is one of the world's better tree climbers. One reason I had come to Humboldt was to see Sillett climb a redwood.

Sillett conducts his research with a group of graduate students and a few other scientists, whom he trains in advanced tree-climbing techniques. I had never met Sillett, and didn't quite know what to expect. "Steve can come across as brusque, but it's because he's so focused," his brother, Scott Sillett, explained to me one day on the telephone. Scott is an ornithologist at the Smithsonian Institution, in Washington, D.C., who studies songbirds. "Steve is one of the most passionate and curious scientists I know. I love birds, but even I can get sick of them. Steve never seems to get sick of trees."

I found Steve Sillett in his office, a windowless room with white walls, stretched out in a chair before a computer, working on some data from the canopy. He was wearing olive-green climbing pants, a pullover shirt, and mud-stained climbing boots. Sillett is thirty-seven years old, of medium height, and he has a lean body with flaring shoulders, huge forearms, and adept-looking hands. His hair is light brown and feathery, and his eyes, set deep in a square face, are dark brown and watchful. Sillett's manner is usually laid-back, but he can act fast. He once fell out of the top of a big redwood named Pleiades I, which is three hundred and ten feet tall. Dropping through the air, he reached out and caught a branch with one hand. This ripped his shoulder out of its socket and tore a bit of flesh out of his hand, but it also stopped his fall. He ended up hanging from the branch by a bleeding hand and a dislocated shoulder, twenty-eight stories above the ground and feeling a bit surprised with himself. "We're primates. Those opposable thumbs are awesome," he explained.

The forest canopies of the earth are realms of unfathomed nature, and they are largely a mystery. They are also disappearing—they are being logged off rapidly, burned away, turned into fragments and patches—and they are perhaps being altered by changes in the earth's climate caused by human activities. Sillett said, "We want to try to understand the basic biology of the redwood canopy, because it will give us a blueprint for how things should work in an old-growth redwood forest."

Redwoods are able to reshape the local climate and environment in which they live. They change the chemical nature of the soil, and they assume control of vital resources in the forest, particularly sunlight and water. "We're trying to get a feel for how much water is stored in the canopy—in the trees, in their foliage, in the canopy soil, and in other plants that live in the canopy," Sillett said. "There's a lot of water up here. These trees are controlling the movement of the water in this forest. How do they do that? What will happen to these forests as the climate changes with global warming? The way the world is going, our work in the canopy could be just a task of documenting something before it winks out."

In 1995 Sillett received a Ph.D. in botany from Oregon State University, in Corvallis. Soon afterward, he took his present job at Humboldt and began to explore the old-growth redwood canopy. No scientist had been there before. The tallest redwoods were regarded as inaccessible towers, shrouded in foliage and almost impossible to climb, since the lowest branches on a redwood can be twenty-five stories above the ground. From the moment he entered redwood space, Steve Sillett began to see things that no one had imagined. The general opinion among biologists at the time—this was just eight years ago—was that the redwood canopy was a so-called "redwood desert" that contained not much more than the branches of redwood trees. Instead, Sillett discovered a lost world above Northern California.

The old-growth redwood-forest canopy, Sillett found, is packed with epiphytes, plants that grow on other plants. They commonly occur on trees in tropical rain forests, but nobody really expected to find them in profusion in Northern California. There are hanging gardens of ferns, in masses that Sillett calls fern mats. The fern mats can weigh tons when they are saturated with rainwater; they are the heaviest masses of epiphytes which have been found in any forest canopy on earth. Layers of earth, called canopy soil, accumulate over the centuries on wide limbs and in the tree's crotches—in places where trunks spring from trunks—and support a variety of animal and plant life. In the crown of a giant redwood named Fangorn, Sillett found a layer of canopy soil that is three feet deep. Near the top of Laurelin, or the Tree of the Sun, which is three hundred and sixty-eight feet tall and still growing, Sillett found a huge, sheared-off trunk with a rotted, damp center. Masses of shrubs are growing out of the wet rot, sending their roots down into Laurelin.

Sillett and his students have found small, pink earthworms of an unidentified species in the beds of soil in the redwoods. A Humboldt colleague of Sillett's named Michael A. Camann has collected aquatic crustaceans called copepods living in the fern mats. The scientists have not yet been able to determine the copepods' species. Sillett said, "They commonly dwell in the gravel in streams around here." He can't explain how they got into the redwood canopy. A former graduate student of Sillett's named James C. Spickler has been studying wandering salamanders in the redwood canopy. The wandering salamander is brown and gold, and it feeds on insects, mainly at night. Spickler found that the salamanders were breeding in the redwood canopy, which suggests that they never visit the ground—this population of salamanders appears to live its entire life cycle in the redwood canopy.

Old redwood trees are infested with thickets of huckleberry bushes. In the fall, Sillett and his colleagues stop and rest inside huckleberry thickets, hundreds of feet above the ground, and gorge on the berries. He and his students have also taken censuses of other shrubs growing in the redwood canopy: currant bushes, elderberry bushes, and salmonberry bushes, which occasionally put out fruit, too. Sillett has discovered small trees—wild bonsai—in the canopy. The species include California bay laurel trees, western hemlocks, Douglas firs, and tan oaks. Sillett once found an eight-foot Sitka spruce growing on the limb of a giant redwood.

OVER THE YEARS, forest-canopy researchers have developed a variety of techniques for gaining access to forest canopies. These include towers, walkways, balloons, high-powered binoculars (the researcher looks at the canopy but doesn't go there), direct tree climbing (which is what Sillett does), and construction cranes. There is, for example, the Wind River Canopy Crane, which is situated in the Gifford Pinchot National Forest, in southwest Washington. The crane has a gondola at the end of a long arm, which can deliver a scientist to any point in the canopy that's within reach, including to the tips of branches, and the gondola can go as high as two hundred and twenty feet above the ground. The disadvantage of a crane is that it's expensive to operate, and it's rooted in one spot. The Wind River crane is able to penetrate roughly one and a half million cubic meters of tree space. Steve Sillett and his colleagues are able to explore roughly twenty billion cubic meters of old-growth redwood forest. In any event, the Wind River crane wouldn't be able to get a person into the crown of the taller giant redwood trees, since many of them are mostly or entirely above the highest reach of the crane.

When Sillett and his colleagues are aloft in redwood space, and moving around from place to place, they make use of tree-climbing ropes and a safety harness called a tree-climbing saddle. They wear helmets and soft-soled boots (spikes can damage a tree). Tree climbers normally hang suspended in midair in a harness attached to ropes looped over solid parts of the tree above them called anchor points. The ropes are attached to the climber's saddle by means of carabiners, which are strong aluminum or steel clips. Tree climbers often move very lightly over branches, keeping most of their weight suspended on their anchor ropes. A skilled tree climber can travel horizontally or at diagonals through the crown of a tree while he's hanging in midair, and not even touch the tree with his body. He may toss a length of rope here or there, getting the rope over a new anchor point, and then he can pull himself to a different place in the tree.

Tree climbing is quite different from rock climbing. Rock climbers move upward over a vertical surface of stone by using their hands and feet to obtain friction and support. They are not suspended from taut ropes (except sometimes in the type of rock climbing called aid climbing). A rock climber advances upward while a safety rope, held by a person called a belayer, trails loosely below him. The rope is there in case the rock climber loses his grip on the stone and falls. In tree climbing, the rope is used as the main tool for gaining height and for moving around. The bark of a tree is crumbly and soft, and a climber can't get any kind of secure grip on it with his hands and feet. The branches of a tree can snap off.

The method by which Steve Sillett and his colleagues climb redwoods is known as a modified arborist-style climbing technique. Arborists, or tree surgeons, get around in trees using a special soft, thick, strong rope. When the rope is passed over an anchor point, it is formed into a long loop or noose, which is tied with a sliding knot called the Blake's hitch. The Blake's hitch was invented in 1990 by a tree surgeon in California named Jason Blake. He popularized it among arborists. It looks a little bit like a hangman's knot, though it functions in a different way. It is a friction knot—it can grip securely on a length of rope, or it can slide on the rope, depending on how the knot is manipulated. By sliding a Blake's hitch, a tree climber can shorten or lengthen a loop of rope over an anchor point, and move upward or downward. A climber can also lock the Blake's hitch, and thereby hang motionless.

In the late 1980s, Sillett was a biology major at Reed College in Portland, Oregon, and had already decided that he wanted to become a canopy scientist; for his senior thesis at Reed, he began climbing tall Douglas-fir trees. A few years later, when he was a graduate student at Oregon State, he got a telephone call from an arborist named Kevin Hillery, who had happened to read Sillett's college thesis. Sillett told him that he was using boots with spikes, called climbing spurs, and a short rope called a flipline to get himself up the trunks of the trees—he was climbing the Douglas firs in the way that an electrical worker climbs up a telephone pole. Hillery said, "If you need spurs to climb a tree, you shouldn't be climbing trees." He offered to teach Sillett the arborist-style climbing technique. Sillett soon began developing his own style of climbing. Sillett's method differs from the classical arborist technique in that he ascends into the crown of a redwood along a single strand of rope called the main rope. Once he gets up into the crown, he detaches himself from the main rope and moves around using a shorter length of rope. (Most arborists don't descend trees on a main rope, since most trees aren't tall enough to require one.)

The process of getting into the crown of a tree is known to tree climbers as the entry. To accomplish an entry into a redwood, Sillett ties a fishing line to an arrow and then, with a hunting bow, shoots the arrow over a strong branch somewhere in the lower crown of the tree. He then ties a nylon cord to one end of the fishing line. By pulling on the other end, he drags the cord over the branch. He ties a rope—which will be the main rope—to the cord and drags it over the branch. Now the main rope is hanging in an upside-down U over a branch in the crown of the tree. He ties one end to a smaller, nearby tree, and then he climbs up the loose, hanging end of the rope using mechanical ascenders (of the kind that rock climbers and cavers use). Sillett says, "You loft yourself into the lower crown on a main rope, and then you detach yourself and move from branch to branch."

To do so, Sillett uses a complicated rig of rope and carabiners, sixty feet long, which he calls a double-ended split-tail lanyard, or a motion lanyard. The lanyard works in the same basic way as does Spider-Man's silk. A climber can attach either end to an anchor point. By attaching alternate ends, he can move around while always staying attached to the tree by at least one strand of rope. Sillett is not immune to a fear of heights. At odd moments when he's aloft in the deep canopy, often when he's hanging from a branch on a motion lanyard, a kind of waking dream flashes over him. For a second, he seems to feel something break, and then he seems to feel himself turning in space as he falls for twenty or thirty stories along the trunk of a giant coast redwood.

The crown of an ancient coast redwood can bristle with rotting extra trunks, and it can be crisscrossed with dead limbs that may be up to several feet in diameter, and there can be broken-off dead branches hanging in the foliage, which are called widow-makers. The twitching movement of a climbing rope can stir loose a widow-maker, and a falling branch can tear off other branches, triggering a cascade of spinning redwood spars the size of railroad ties. A falling branch can spike itself five feet into the ground. Redwoods can have pieces of dead wood in them that are bigger than Chevrolet Suburbans. Sillett carries a little folding saw with him and uses it to cut away any small, hazardous dead branches, "but with the big hazards we just have to rely on hope," he says. Redwoods occasionally shed whole sections of themselves. Sillett calls this process calving. The tree releases a kind of woodberg, and as it collapses it gives off a roar that can be heard for a mile or two, and it leaves the area around the calved redwood looking as if a tank battle had been fought there. A calving event would obliterate any humans in the tree. Sillett told me, "The thing I fear most is a falling branch that hooks on my rope. It would slide down the rope into me, and it would tear through my body cavity. You are a grape hanging on a vine, and a falling branch can pop you."

The tallest living part of a tree is called the leader. The leader of a coast redwood is often a delicate spindle, covered with papery bark, and its branches are brittle. When Sillett reaches one, he sometimes takes off his boots and socks and hangs them on a branch, and then he climbs barefoot to the top. "It makes sense to have a form of communication with the tree," he says.

In the summer of 2000, while Sillett was attending a scientific conference, he met a canopy scientist named Marie Antoine. They started climbing redwoods together, and were married soon afterward. Antoine, a slender woman in her late twenties, was born and raised in Canada. She is an expert in lichens and a lecturer at Humboldt State. Sillett and Antoine live in a small yellow house in the hills overlooking Arcata. They keep their tree-climbing gear in the garage.

IN THE 1840S WHEN AMERICAN SETTLERS arrived in Northern California, the redwood forest amounted to roughly two million acres of virgin, old-growth trees. Loggers began cutting down the redwoods with axes and handsaws, using the wood for making barns, houses, fences, and railroad ties. In the 1920s and 1930s, the introduction of logging machinery, chainsaws, and Caterpillar tractors vastly increased the speed of logging along the northern coast of California, and the old-growth redwood forests began to disappear. Most of these forests ended up being owned by timber companies. As a rule, they carried out clear-cutting operations, in which no tree of any worth was left standing. In all, close to 96 percent of the virgin redwood forest was cut down. Some botanists, including Steve Sillett, believe that during the logging a number of redwood titans were felled that were bigger than any of the living giant sequoia trees are today. In other words, before the logging, the coast redwood was probably the largest tree on earth, not just the tallest.

About ninety thousand acres of old-growth redwoods have remained intact, in patches of protected land. The remaining scraps of the primeval redwood-forest canopy are like three or four fragments of a rose window in a cathedral, and the rest of the window has been smashed and swept away. "Oh, man, the trees that were lost here," Sillett said to me one day as we were driving through the suburbs of Arcata. "This was the most beautiful forest on the planet, and it's almost totally gone. This is such a sore point."

Conservationists won a major victory in 1999, with the signing of a deal known as the Headwaters Agreement, between the federal government and one of the largest redwood-timber companies, Pacific Lumber, which is a subsidiary of Maxxam. The agreement, and subsequent deals, gave the government title to a large part of Pacific Lumber's old-growth redwood tracts, including the seventy-five-hundred-acre Headwaters Forest, half of it virgin redwood, in the mountains southeast of Eureka, California. The government paid Pacific Lumber $380 million. However, stretches of virgin redwood forest left on private timber-company lands continued to be logged.

Today, the timber-company lands in Northern California—owned by Pacific Lumber, Green Diamond Resource Company, and others—

are managed for high-volume production of young redwood trees. The tracts are logged off on a schedule, typically every fifty years or so. It is increasingly difficult to find any redwood trees growing on timber-company land which are more than eighty years old. Sillett has climbed in these trees. Their crowns are nearly devoid of life. They are a redwood desert.

In his office, on the day I first met him, Sillett tapped on a mouse, and a computer-generated image of a grove of redwoods appeared on the screen. "You need to look at some redwood architecture," he said. The grove, he explained, is named the Atlas Grove, and it is in Prairie Creek Redwoods State Park, which is a sliver of protected old-growth redwood forest that hugs the sea along the northern part of Humboldt County. The computer image was a three-dimensional map of the crowns of the trees in the Atlas Grove; he had made the map from data that he and his associates had gathered in years of climbing them. The Atlas Grove is tiny—it is about three hundred and fifty yards long by thirty-five yards wide—and it is a jam of monstrous redwoods with reiterated crowns. The Atlas Grove may be the oldest grove of living redwoods; although Sillett isn't sure exactly how old it is, judging by the height, the amount of rot in the dead parts of the trees, and the types and abundance of epiphytes, it seems to have come into existence around the time of Julius Caesar. The Atlas Grove was discovered in 1991 by a tall-tree explorer named Michael Taylor, who is a friend of Sillett's. Discovery, in the case of a giant tree, doesn't necessarily mean that the tree has never been seen before—the grove had been known to some park rangers. It means that nobody has understood the tree's size, or has measured it.

Sillett named most of the trees in the grove after Greek gods. (Generally, the discoverer of a redwood names it: Sillett often gives the tree a name only after he's climbed it and gained a sense of its character. There's no formal process—the names are private, known mostly to the botanists who study redwoods.) "That's Prometheus, that's

Epimetheus, and that's Atlas—they were all brothers," Sillett said, pointing to the screen. The trees looked something like witches' brooms standing on their handles. He tapped the mouse again, and the images began to rotate. "Atlas has these three huge trunks in its top, like a trident," he said. "Atlas is so full of soil and plants that you get this overwhelming sense of a tree holding up the earth. Here's the Pleiades. And that's Kronos and that's Rhea."

The biggest tree in the Atlas Grove is a redwood titan named Iluvatar. Sillett named the tree after the creator of the universe in J. R. R. Tolkien's *The Silmarillion.* "Iluvatar is so complex that you can't tell much about it just by looking at it," he said. On the screen, Iluvatar rotated slowly, as if we were flying around it in a helicopter. The crown of Iluvatar contains two hundred and ten trunks, and it fills thirty-two thousand cubic yards of space. It took Steve Sillett, Marie Antoine, and four graduate students roughly ten days of climbing apiece to make a 3-D map of the crown. With these data, Sillett, along with an expert in giant trees named Robert Van Pelt, performed a calculation that shows that Iluvatar contains thirty-seven thousand five hundred cubic feet of wood. Iluvatar is one of the largest living things on the planet.

I asked Sillett where the Atlas Grove was, exactly.

He gave me a guarded look. "It's something you can't print."

Botanists are secretive about the locations of rare plants. They fear that any contact between humans and rare plants can be disastrous for the plants. Sillett is particularly worried about recreational tree climbers. Recreational tree climbing is an evolving sport, or emerging oddity, which is practiced by a thousand or so people in the United States but is rapidly growing in popularity. It apparently got its start in 1983, when a certified arborist in Atlanta named Peter Jenkins began teaching all sorts of people, including children, how to climb trees safely using a rope and a harness and the arborist climbing technique. He founded a tree-climbing school called Tree Climbers International. The classroom of Tree Climbers International consists of two white-oak trees on a plot of land near downtown Atlanta.

Sillett is widely admired by recreational tree climbers, but his feelings about them are laced with foreboding. "Not only are redwoods sensitive to damage from climbing but the whole habitat of the redwood canopy is fragile," Sillett explained, that day in his office. "It cannot survive without damage if people start climbing around in it for recreational reasons." He believes that recreational climbers would try to climb the biggest and tallest redwoods if they knew their exact locations, and they wouldn't bother to get anyone's permission. Climbing a tree without permission is an accepted part of the culture of recreational tree climbing. It is called doing a ninja climb or poaching a tree.

The U.S. National Park Service is in charge of issuing research permits for the parks where most of the remaining giants and titans among the coast redwoods stand. Sillett and other canopy scientists are allowed to climb in the parks only between the middle of September and the end of January each year. During the rest of the year, the National Park Service closes the redwoods to climbing in order to allow the spotted owl and a seabird called the marbled murrelet to nest in the redwoods without being disturbed.

Sillett closed the view of the Atlas Grove on his computer, and stood up. "So, have you seen enough?"

"Not really. I was wondering if I could climb in the redwood canopy with you."

He didn't answer. He looked me over with a kind of professional coolness. His eyes seemed to focus on my face and neck, my torso, and my hands. "Are you a tree climber?"

"Yes, I am."

"These trees are gnarly. All it takes to get yourself killed is one mistake."

I told him that I knew something about climbing trees. I'd come across the Tree Climbers International school one day while I was surfing on the Internet. I had never really thought about tree climbing; it sounded weirdly appealing. I got a flight to Atlanta, and I began to learn the art of movement in a forest canopy. By the time I met Steve Sillett, I had had about twenty hours of basic training. In addition, I had climbed half a dozen trees near my house, which is in New Jersey. But, in Sillett's view, I knew essentially nothing about climbing redwoods. I had never heard of some of the advanced equipment that Sillett uses, like the motion lanyard; in fact, many professional tree climbers have never heard of it, either.

Sillett said that while he was happy to have me walk around on the ground in the redwoods with him and his colleagues, he was not interested in taking me up into the canopy, at least not immediately. (His brother, Scott, who climbs with him every now and then, explained later, "It's definitely true that Steve doesn't trust the climbing skills of others. It takes such focus to climb safely.")

And so I returned to Atlanta for further training. One day in April, I found myself seventy feet above the ground, dangling from a branch on a rope like a Christmas ornament, practicing the skills needed to get oneself safely from place to place in a tree. It was a cool, blue day, and the wind was blowing. I was suspended at the point of a V formed by two loops of ropes. By sliding Blake's hitches, I shortened one loop while I lengthened the other. This changed the shape of the V, from an asymmetrical V oriented toward the left to an asymmetrical V oriented toward the right. In this way, I traveled horizontally in midair to a different place in the tree. In the distance, the rectilinear towers of Atlanta glittered in the sun.

Back home in New Jersey, I ordered from an arborist supply company sixty feet of half-inch tree-climbing rope, four carabiners, and two split tails, which are short pieces of rope that are tied with Blake's hitches to the longer piece of rope. When these things arrived, I assembled a motion lanyard. At first, I practiced with the lanyard while I was standing on the ground—I heaved its ends up over branches in a maple tree in my front yard. Then, wearing a climbing saddle and a helmet, I raised myself into the air with the lanyard, and got about six feet off the ground. In an ash tree that grows off to the side of my house, I ascended sixty feet by throwing the ends of the motion lanyard over higher and higher anchor points—that is, I lanyarded my way to the top of the tree. I extended my circle of climbing, and began to explore a forest canopy that ranges up a hillside above my house.

The forest canopy on my hill extends from about fifty to a hundred feet above the ground. It is composed of the crowns of sugar maples, red oaks, white oaks, chestnut oaks, hickories, ash trees, tulip poplars, and some tall, beautiful old beech trees. As I became more adept at movement in trees, I became better able to go laterally, or on diagonals, through the air. Birds seem to pay little attention to a person hanging on a rope in a forest, and it's not always clear that they are able to identify such an object as a human being. Sometimes flocks of birds sweep through the canopy and divide around a climber or move beneath him. Flying squirrels are tame in the canopy. I've reached out a finger and stroked their fur on occasion. (They close their eyes when you stroke them, but they soon tire of it, and plummet off the branch, catch the air, and soar away.) During climbs into taller trees, I was occasionally able to look down on the backs of birds, which shine with reflected sunlight as they move through the green depths of the canopy, like schools of fish.

SILLETT AND GEORGE W. KOCH, a tree physiologist from Northern Arizona University, and a graduate student named Anthony R. Ambrose have installed electronic monitoring systems in seven of the biggest and tallest redwoods. The monitoring systems consist of weather stations at various spots in the trees, and different kinds of bio-probes, and the systems are linked together with up to half a mile of data cables strung in each tree. The scientists are trying to learn how redwoods move water through their trunks and branches and how they manage to grow so tall.

George Koch is a lanky, genial man in his forties, with knotted arms and an easygoing manner. Sillett taught him how to climb trees. "I'm like a kid in a candy shop, climbing these three-hundred-and-sixty-foot-tall trees with Steve," Koch said to me one day when I visited him in Arcata. "The overwhelming question for me is what determines the height of a tree. At around three hundred and seventy feet, the tallest redwoods seem to be approaching a ceiling, which is based on a limit in height to which any plant can lift water. Why aren't the redwoods six hundred feet tall?"

Exactly how an extremely tall tree delivers water to its top is a matter of deep interest to Koch and Sillett. Trees bring water upward from their roots through a network of microscopic pipes called the xylem. The pipes are unbroken from the bottom to the top of a redwood. It takes a few weeks for water that is absorbed in a redwood's roots to get to the top of the tree. A redwood can move water upward through its pipes against two million pascals of negative pressure. (The pascal is a standard measure of pressure used by physicists, and negative pressure is basically a sucking force.) Sillett and Koch have been looking for an engineered system that sucks water at two million pascals, so that they can do experiments. But they can't find such a system. Apparently, redwoods are better at pulling water than any human technology is.

In a spell of dry years, air bubbles seem to form in a redwood's pipes and stop the continuous flow of water, and the top dies and usually falls off. In wetter periods, a redwood regrows its top. "Redwoods have an incredible ability to reiterate new trunks," Koch went on. "A side branch will take off and shoot skyward, and in a matter of a hundred years it will become the new leader—the new top—of the tree."

In any ecosystem in which they occur, redwoods tend to dominate. They tower above other species of trees, and they shade them out, killing them or making it nearly impossible for them to grow. Trees are horrible to one another, and redwoods are viciously aggressive. They drop large pieces of dead wood on smaller neighboring trees, which typically shatters the tree. Sillett calls this phenomenon "redwood bombing." In this way, a giant redwood suppresses and kills trees growing near it, including hemlocks, spruces, Douglas firs, and big-leaf maple trees. A giant redwood can clear a DMZ around its base, an area covered with redwood debris mixed with twisted and dead trees of other species.

Redwoods are monoecious, which means that the plant is both

male and female. The female organs of the tree are its round, knobby seed cones. A redwood's male organs are small, nubbin-like cones, called strobili, which appear at the tips of branches and release pollen. The grains of pollen contain sperm cells that fertilize egg cells inside the female cones, and seeds are produced. Over its lifetime, a redwood can release ten billion seeds. It may be that only one of the seeds gets lucky and becomes a mature redwood tree. If a redwood is sheared off at its base, or if it burns to a blackened spar, it can send up from its roots a circle of small trees, or clones, having DNA identical to the parent tree's. The clones become a ring of redwoods in the forest, forming a structure called a cathedral tree.

Redwoods are exceedingly efficient at gathering light. A grove of redwoods is able to soak up more than 90 percent of the sunlight falling on the crowns of the trees. Young redwoods are able to survive in dark places, where almost no other trees can survive, since they come into existence in the deep shade of their elders. When an old redwood falls, creating an opening in the canopy, sunlight splashes onto smaller redwoods, and they leap upward, rapidly becoming big trees. Little redwoods can sometimes crop up in thickets of slender trunks.

Scientists suspect that such a group of small redwoods may be joined at its roots, and also may share a common root system with a large redwood nearby—but nobody really knows. The small redwoods and the big redwood may all exchange water and nutrients with one another. It is possible that the root systems of the redwoods in a forest are fused into a web underground, so that they can be thought of as a single living thing. These are all questions that remain unanswered by science.

IN THE DARKNESS BEFORE DAWN on a cold November morning near Arcata, Marie Antoine was hurrying around her kitchen. She was singing to herself in a dreamy kind of way. That morning, she wore a gray hooded cashmere sweater, cream-colored slacks, and climbing boots. She tossed a handful of blueberries into a blender. "Steve, do you want a smoothie?"

"Definitely," he answered. He and I were kneeling on the living-room floor, nearby, and he was inspecting a heap of my climbing equipment, which I'd taken out of a duffelbag. He stood up, and led me through a door into the garage, where a hank of climbing rope, sixty feet long, was coiled on a hook on the wall. It was a new rope, and it was bare—it didn't have any carabiners or knots rigged in it. "This is yours, if you want it. I'll tie it up for you."

"I brought my own split tails," I said. I carried the rope back into the living room, and laid it on the floor, and from the duffelbag I took out two split tails and four carabiners, and I tied up my motion lanyard while he watched.

He sipped his smoothie. "Dude, you're doing it. Where did you learn this?"

"I've been practicing a little," I said. I tied the last Blake's hitch and sat back on my heels.

He inspected the knots. "Sweet," he said. He picked up the motion lanyard and handed it to me. "You can stuff it into your climbing bag." A motion lanyard weighs eight pounds. When it isn't in use, it is kept inside a bag, which is normally clipped to the climber's foot stirrup.

The blender whirred. "Which tree are we going to?" Antoine asked.

"I think we need to go to Adventure," Sillett said.

Adventure is one of the world's largest-known coast redwoods. It is three hundred and thirty-four feet tall, and it contains thirty-one thousand cubic feet of wood. The main trunk is sixteen feet in diameter near ground level, and it maintains huge girth nearly all the way to the top of the tree. It has a total of forty extra trunks. Adventure has four fire caves in its crown, two of which are large enough for a person to go inside. Much of the center of the tree seems to be rotten. It is in Prairie Creek Redwoods State Park, a few miles from the Atlas Grove. The precise location of Adventure is a secret that is known to fewer than twenty people, most of whom are botanists. Sillett and

Antoine asked me not to reveal its exact location, for fear that recreational tree climbers might try to poach it.

"Adventure Tree is never exactly my first choice," Marie Antoine commented. "My first experience climbing that tree was kind of scary."

Later, I asked her what had been scary about Adventure.

"I got lost in it."

THE STRETCH OF THE CALIFORNIA COAST that includes Prairie Creek Redwoods State Park is covered with temperate rain forest; it receives eighty inches of rain a year. Even so, places in the park consist of patches of open prairie, where herds of elk graze. The redwoods along the edge of the prairie looked like ruined Doric columns. The road went among them, the canopy closed in overhead, and the world grew dark and quiet—redwoods mute sounds. Adventure lives at the bottom of a small valley. We parked and put on backpacks full of tree-climbing gear. Marie Antoine led the way. We went along a trail, and then left the trail and pushed through masses of sword ferns and walked in zigzags around them. The ferns were chest-high and were soaking wet—it had rained during the night. The forest floor consisted of a soft duff of rotting redwood foliage, and it was spattered with redwood sorrel—small, emerald-green plants that have heart-shaped leaves. The trunks of the trees soared into remote crowns. Blades of sunlight angled through the canopy, and they glittered with droplets of water falling from the tips of branches. The sky was pale blue, without a cloud.

We went down into a gully and arrived at a small creek. A redwood log spanned the creek, forming a natural bridge across it. Adventure grew out of the bank on the other side: I saw a megacylinder of wood with a thermonuclear crown.

We crossed the creek by walking on top of the log, which was slippery, and scrambled up onto the bank at the base of the tree. Sillett opened his backpack and pulled a climbing rope from it. He threaded the rope up through an attachment point higher in the tree and back down to the ground.

The rope was six hundred feet long, dusty black in color, and just ten millimeters thick. It had a breaking strength of three tons—it was strong enough to lift a car. It is sometimes called black tactical rope, and it is favored by the Special Forces for vertical operations at night. The redwood scientists like it because it's lightweight yet super-strong and can seem nearly invisible. "We like to go low-pro," Sillett explained.

The black rope hung down the side of the tree which faced away from the stream, a nearly featureless shaft without any solid branches on it for two hundred and fifty feet. Sillett took one end and tied it to a small tree. The other end dangled loosely down from the anchor point, far up in the crown; he would climb up the dangling end of the rope. This is known as a ground-anchored climbing rope. He put on his helmet and his climbing saddle. He turned on a two-way radio, tested it, and put it in a pocket on his chest.

While Sillett was getting ready to climb, Antoine led me around to the other side of the tree. It consisted of a towering system of extra trunks, some living and some dead, that ran upward along the stream side of Adventure for more than twenty stories. Antoine put her hands in her pockets and stared up into the structure. "The pieces of dead wood shiver fifty feet over your head when you move around in there," she said, peering up into it. "The first time I climbed this tree with Steve, he told me to go to a certain place, and I misunderstood." She ended up wandering among columns of rotten wood, which wobbled and seemed ready to collapse. Finally, she tied herself to a live branch and called her husband on the radio and asked him to come and get her and show her the way out. It took him twenty minutes to find her, and when he did she was embarrassed—she didn't like needing to be rescued.

Back at the other side of the tree, Sillett clamped a pair of mechanical ascenders to the black rope, and then he began to climb upward, sliding his ascenders on the rope in a one-two type of motion which climbers call jugging. A raven called somewhere in the upper canopy. This was followed by a delicate *pip, pip, pip,* which was coming from somewhere closer to the ground.

Antoine picked up her radio. "Steve, was that a kinglet?"

"It could be a wren tit." He had become a tiny homunculus moving up the bulwark of the tree.

"Adventure just scoffs at the puny humans who try to climb it," Antoine remarked to me.

Sillett vanished into the crown. Time passed. We put on our helmets and saddles. I sat down under a fern, and picked up a sprig of dead redwood foliage. Two mushrooms grew on it.

Our radios crackled. "Okay, Marie, you can go ahead and release the anchor."

"Okay, I'm going to untie the anchor, Steve. I will let you know when it's done." Their radio talk was precise. Any mistake in communication could result in someone's death. (Sillett once saw a climbing companion fall ninety-six feet from a Douglas fir. The friend was a professional climber who had made one small mistake. Miraculously, he lived.) Antoine went over to the small tree and untied the knot. "The anchor is now untied, Steve."

Sillett, up in the tree, tied a knot near the middle of the black rope, and then anchored the knot around a branch, so that both ends of the black rope hung down along different sides of the tree. Antoine and I would simultaneously climb up the opposite ends of the rope.

"You guys can start moving up," Sillett said on the radio.

I clamped my ascenders to my length of the black rope, and I clipped the bag containing the motion lanyard to my foot stirrup. Then I began to jug upward on the rope, along the basal flare of the tree. Twenty feet above the ground, the tree's bark was charred and pitted with fire scars—a small fire on the ground had made the scars, perhaps within the past two hundred years. I kept on jugging. Seventy feet above the ground, I passed a burl, which is a type of benign growth that occurs on trunks. The burl was the size of a pumpkin. I continued climbing upward, along a furrowed wall of wood. Marie

Antoine was climbing somewhere around the horizon of the trunk, out of sight.

I reached a height of about a hundred and thirty feet. I was now forty feet higher than I had ever climbed in the canopy in New Jersey, and I was just entering the lower edge of the redwood crown. The light began to brighten. The rope to which I was attached ran straight upward into the crown, and it vibrated with tension from the weight of my body. When canopy scientists want to travel in a circle around the trunk of a large tree, they swing like a pendulum at the end of a long rope. I decided to try it. I planted my feet against the tree, and pushed off. I drifted a considerable distance outward, floating gently away from the trunk. On the forest floor below, the clumps of sword ferns looked like green stars. They turned around—I was spinning. I drifted back to the tree, and kicked off again, harder, and drifted farther away from the tree. It seemed, perhaps, like walking on an asteroid, where there is only slight gravity. The curve of the trunk formed a horizon like that of the small worlds in *The Little Prince.*

Marie Antoine appeared—she had circled around the tree to see how I was doing.

The bark was covered with a lumpy white crust that looked like sugar frosting. "What's this stuff?" I asked.

"It's Pertusaria. It's a lichen."

Pertusaria is also called wart lichen. I moved my eyes closer. The warts were mingled with splotches of a grayish-green dust, which was sticking to the bark. The dust, Antoine explained, is another lichen, called Lepraria. It is supposed to resemble the infected skin of a leper. The Lepraria, in turn, was mixed with fingering spurts of a lichen called Cladonia. The Cladonias are among the most beautiful of lichens. They come in wild shapes—trumpets, javelins, stalks of pinto beans, blobs of foam, cups, bones, clouds, and red-capped British soldiers. This Cladonia looked like pale-green tongues of flame. Scattered near it were clusters of orange disks that looked like tiny pumpkin pies. It was a lichen called Ochrolechia, Antoine said. The cracks in the bark were lined with pin lichens—tiny black dots standing up on stalks, like the heads of pins shoved into the wood. It occurred to me that in order to see a giant tree you need a magnifying glass.

We climbed side by side for a distance, until we arrived at a stob—a dead, broken-off stump of a branch. It was surrounded by huckleberry bushes and leatherleaf ferns. The ferns trembled in a breeze that flowed up the side of the tree. The stob was home to what looked like a miniature Japanese garden, about six inches across. We hung suspended in the air before the garden, while Antoine pointed out its sights: "That's Lepidozia, a liverwort. That's a little liverwort called Scapania—it looks like tiny ladders." She pointed to tufts of shimmering, bright-green moss. "That's Dicranum. It's all over redwoods." She estimated that the garden on the stob was several hundred years old.

I climbed for a while on the other side of the tree. I wanted to see the array of trunks which looms over the creek, on the dark side of Adventure. I pendulated in that direction. When I arrived there, I found myself in the middle of a Gothic tower of fusions, bridges, and spires, held up by flying buttresses. The zone was crisscrossed with branches, and the trunks ran out of sight in both directions, upward and downward.

Directly in front of me, at a height of a hundred and eighty feet, was a fire cave. It is called the Upper Fire Cave, and its mouth is plastered with dirt—canopy soil. By gripping on ridges of bark, I was able to pull myself to the edge of the cave. It proved to be a sort of airy chamber in the underside of a flying buttress, and it opened downward into empty space—it is more like a fire ceiling than like a fire cave. I ended up hanging in midair, a few feet below the charred ceiling, looking straight down to the stream. There was a faint sound of rushing water. I saw strands of computer cables emerging from the cave walls, where Sillett and his team had implanted electronic probes. I touched the wall of the cave. It was moist, and it had a yellowish color and a musty smell, and it felt like Stilton cheese.

Two hundred and fifty feet up, the crown of Adventure billowed into a riot of living branches. By this time, the ground had disappeared—it was hidden below decks of foliage in the lower parts of the redwood canopy—and the sky was invisible, screened by tents of foliage overhead. This was the deep canopy, a world between earth and air. Steve Sillett was nowhere to be seen.

I was now climbing fifteen feet above Marie Antoine. Even though we were still far below the top of Adventure, we were considerably higher than the top of the average tropical-rain-forest canopy. If the upper surface of the Amazonian canopy had existed here, it would have been a hundred feet below us. At this point, the main trunk of Adventure was seven feet in diameter. Now the huckleberry thickets began in earnest. The species was the evergreen huckleberry, a relative of the wild blueberry that grows in Maine. In November, in the California rain forest, the huckleberry leaves were tinged with scarlet at their edges. The bushes were all over the tree: perched on its branches, occupying its crotches, and popping out of the trunk. I wormed through them, following the black rope upward.

AT TWO HUNDRED AND NINETY FEET, I encountered Sillett. He was sitting on a branch inside a spray of huckleberry bushes, and he had a thoughtful look on his face. The main trunk had split open near the branch where he sat, and the opening revealed dead and rotten wood inside the tree. "This beast is full of rot pockets," he said. "These huckleberry bushes are putting their roots through the scars into rotten wood in the center of the tree. One summer, we had half the normal rainfall, but these bushes still put out a full crop of huckleberries. They're getting their water from rotten wood inside the tree." He pointed to something on the side of the tree. "Check out that little brown moss over there."

"Which moss?"

"The one that looks like it's dead."

I hung out over a branch and looked at the moss. It was a greasy-looking thatch growing below a wound at the base of a stob. Redwood sap had been dribbling over the moss.

"It's called *Orthodontium gracile*," he said. "It's an extremely rare moss. It often lives below wounds in old-growth redwoods. It likes the resin. It's nearly gone in Oregon. That's because old-growth redwoods in Oregon have been slaughtered."

I had reached the upper end of the black rope. Nearby, I saw the bottom end of a second climbing rope, a white one, which was hanging down along the trunk of the tree. It wandered upward and out of sight, toward the top of Adventure. Sillett suggested that I get on. I transferred my ascenders to the white rope, and climbed up it for about thirty feet, wriggling through a jungle gym of redwood branches and huckleberry shrubs. The bag that held my motion lanyard bumped along through the bushes. Then, abruptly, the crown thinned out, and a view opened across Adventure Valley.

The white rope came to an end about fifteen feet below the top of the tree. No ropes led to the top. I took the motion lanyard out of my bag, attached it to my saddle with carabiners, and threw one end over a branch above me. I pulled it back to me, to form a noose over the branch, and clipped the noose to my saddle. Then I detached myself from the white rope.

There is something unnerving about leaving the main rope behind and going into free motion in the crown of a redwood tree. The main climbing rope is a lifeline that connects a climber to the ground, and it is the escape route out of the tree. Once you disconnect from the main rope, you are on your own. If you wander far from the main rope, you can end up moving through a maze of wood as tall as an office building by means of a short piece of rope, with no way to get down to the ground unless you can find the main rope again.

With my weight on the motion lanyard, I leaned back, until my body was horizontal and my feet were planted on the trunk, and I walked up the trunk of Adventure. I threw one end of my lanyard over a higher branch, clipped it back to my saddle, and pulled myself up. Suddenly, I hung near the top of the tree. At three hundred and twenty-eight feet, I found myself in the middle of a bush studded with huckleberries. I began eating them. They were tart and crunchy. The branches in the tree's top were festooned with beard lichens—

they looked like the frizzy beards of dwarves. It was a sunny day, and a breeze was blowing, which stirred the lichen beards, and the air held a tang of the sea—the Pacific Ocean lay over a ridge to the west. Adventure rocked in the breeze, like a ship riding at anchor.

The uttermost top of Adventure is dead. It is a gray trunk, encrusted with lichens, which extends about six feet above the huckleberry bush, and ends at a sheared-off stump. Adventure used to be a taller tree. Its top fell off, probably in a storm, perhaps four hundred years ago, or roughly at the time that Shakespeare wrote *The Tempest.* By then, it had already been growing for a thousand years, or maybe more like fifteen hundred years. ("Who the hell knows how old it is," Sillett said.)

The branches around me trembled. A lanyard flipped over a nearby branch, and Marie Antoine appeared. She trunk-walked up to a kind of platform of branches, and sat in the middle of them. "The top of this tree is just a big old juicy dead-wood pit," she said.

The dead trunk at the top of Adventure is a natural water tank, she explained. Rainwater collects in the broken stump at the top, and the water runs down inside Adventure, where it saturates the rotten wood like a sponge. A coast redwood tree seems to have the ability to send out roots from any part of its tissue, including its top. Adventure may be sending roots out of the living wood in its top, which run into the dead trunk, and feeding on the dead parts of itself.

THE NEXT DAY, in the afternoon, I sat down at the base of Adventure, while Sillett went aloft. He had been having trouble with a computer system that he had installed in Adventure, and he wanted to try to fix it. It was getting to be late, and it grew dark on the forest floor, but there was sunlight in the crown of Adventure. I called Sillett on the radio. "Where are you?"

"At the Upper Huckleberry Cave."

"Where's that?"

No answer. I put on my helmet and saddle, and began jugging up the black rope. I found him at two hundred and thirty feet, hanging from his motion lanyard inside a rampart of extra trunks, a long way from the main rope. I climbed up the main rope until I was fifteen feet above Sillett, and then I pulled out my lanyard, flipped one end of it over a branch, and clipped it back to my saddle. Then I released myself from the main rope and dangled on the lanyard. The trunk was a fissured wall going down into shadows. Suddenly, I was very aware that I was hanging more than twenty stories above the ground. I lengthened the noose, and dropped twenty-five feet, until I was hanging below Sillett. I kicked against the trunk, swung away from the tree, and then came back toward him. I grabbed the end of a rope that Sillett had left draped over a branch and, using my ascenders, climbed up the rope at a diagonal to Sillett's location. When I got there, I anchored my motion lanyard over a branch and I hung in the air next to him.

We were suspended below a cracked and decayed expanse of holes in the tree—the Upper Huckleberry Cave. Sillett was hanging in front of a fiberglass box. The box, which was attached to a branch, held a computer controller that gathered data from all the instruments that were installed in the tree. A laptop computer was sitting on top of the box, and he was staring at the screen.

He seemed exasperated. "Every time we climb Adventure, it kicks our butts," he said. "I think it's cursed." He fished a Leatherman tool out of a pouch and began tinkering with it inside the box. The sun had set behind a ridge, and an evening breeze had sprung up. The tree's branches and needles began to give off a hissing sound. "Do you feel that?" he said. "We're moving."

Adventure began to do something that felt like slow breathing. Each sway of the tree took several seconds to complete. The redwoods around Adventure were also tracing deep, slow sways, and their movements were independent of one another: they were going in different directions. The trees seemed intensely alive.

Sillett watched the motion of the redwoods in silence for a little while. "Despite the difficulty of doing science in these trees, there's always a moment during a climb when you can lose yourself," he went on. "You perceive time more clearly in redwoods. You see time's illusory qualities. When you get up into the crown of a redwood, you stop thinking about your life, you stop planning your future missions. You start feeling the limits of your perceptions of the world as a member of the human species. When you feel one of these trees moving, you get a sense of it as an individual."

"Do you really think of this tree as a kind of entity?" I asked.

"It's a being. It's a 'person,' from a plant's point of view. Plants are very different from animals, but they begin life with a sperm and an egg, the same way we do. This organism has stood on this spot for as many as two thousand years. Trees can't move, so they have to figure out how to deal with all of the things that can come and hurt them. This tree has burned at least once. The fire must have continued inside some of these caves for a long time—the caves were smoldering orange holes in the tree for weeks. Redwoods don't care if they burn. After the fire, the tree went, 'Wooaah,' and it just grew back."

The wind died, and the forest became silent. A fluting call came from the air near us. Sillett looked around. "Maybe a Swainson's thrush." He poked with his Leatherman at the electronics. "A tree is not conscious, the way we are, but a tree has a perfect memory. If you injure a tree, its cambium—its living wood—will respond, and the tree will grow differently in response to the injury. The trunk of a tree continually records everything that happens to it. But these trees have no voice. My life's work is to speak for these trees." He paused. "Dude, it's getting dark. It's time to go down."

Three hundred and fifty-three feet above an alluvial flat in the Humboldt Redwoods State Park, near the top of a coast redwood named Idril, Steve Sillett and Marie Antoine were sitting side by side in the branches. I was attached to the tree with my motion lanyard, near them. My lanyard was anchored around the uttermost top of the tree. Idril is about eight inches thick at that point. The top of the tree is crowded with wires and instruments and a solar panel—Sillett's gadgetry. We were at the upper surface of the world's tallest forest canopy.

I tightened my lanyard, and inched myself up a little higher. The whole top swayed.

"Watch your foot," Sillett said. I had nearly kicked a bioprobe out of the bark.

The upper surface of the canopy was a bubbly froth of redwood crowns, and each tree seemed to have a slightly different color of green. There were deep greens, gray-yellow greens, brown greens, deep-blue greens, and bluish grays. "It's because these trees have a huge amount of genetic variability," Antoine said. It began to rain, and the colors grew sharper. Here and there, skeletons of dead tops seemed to glow—these were redwoods entering middle age. "You can see the Stratosphere Giant," Sillett said, pointing to a green cloud that burst above the canopy to the east of Idril.

On July 30, 2000, an amateur redwood researcher named Chris K. Atkins was bushwhacking around in a little-visited stand of redwoods in the park, using a laser range finder to measure the heights of the trees. On that occasion, Atkins discovered what is currently believed to be the world's tallest tree. He named it the Stratosphere Giant. Sillett measures the Stratosphere Giant once a year, each September, when he climbs it with Marie Antoine, and they run a tape measure along the trunk. This past September, they found that the Stratosphere Giant was three hundred and seventy feet two inches tall. It is the only living tree that is known to have surpassed three hundred and seventy feet, and it is presently growing taller by about four inches a year. It is eighteen feet across at the base, and is around two thousand years old.

The rain became steady. We descended Idril in stages, one by one, rappelling down a rope. Partway down, we stopped in a cluster of branches, and we left the main rope and anchored off on our lanyards. With his lanyard anchored around a branch above him, Sillett stepped out onto a branch, lengthening the lanyard as he went. He was using the lanyard to maintain his balance and to prevent his full weight from pressing on the branch itself—this is called branch walking. Sillett walked lightly along the slender branch nearly to its tip. He looked almost weightless, and he leaned out and touched the branches of a neighboring redwood. "You could easily get into that tree from here," he said.

We were now at two hundred and fifty feet, near the bottom surface of the canopy in this grove. I clamped a descender to the black, main rope. I released my motion lanyard and tucked it away in its bag. With my weight now on the main rope, I released the descender brake and began to slide down. Then I kicked away from Idril as hard as I could. As I swung from the tree, I opened the brake on the descender full wide. The rope began to rush through the descender, and I fell out of the canopy on a fast rappel. Huge columns appeared, the trunks of trees that stand around Idril, and I floated weightless down through redwood space.

FRANS B. M. DE WAAL

We're All Machiavellians

FROM THE *CHRONICLE OF HIGHER EDUCATION*

When he was a young primatologist, Frans B. M. de Waal quickly learned that the textbooks he had studied did not help him in understanding the behavior of chimpanzees. Observing chimp society, he soon found that Machiavelli was a better guide for what he was seeing.

Given the obvious "will to power" (as Friedrich Nietzsche called it) of the human race, the enormous energy put into its expression, the early emergence of hierarchies among children, and the childlike devastation of grown men who tumble from the top, I'm puzzled by the taboo with which our society surrounds this issue. Most psychology textbooks do not even mention power and dominance, except in relation to abusive relationships. Everyone seems in denial.

In one study on the power motive, corporate managers were asked about their relationship with power. They did acknowledge the existence of a lust for power but never applied it to themselves. They enjoyed responsibility, prestige, and authority. The power grabbers were *other* men.

Political candidates are equally reluctant. They sell themselves as public servants, only in it to fix the economy or improve education. Have you ever heard a candidate admit he wants power? Obviously, the word "servant" is doublespeak: Does anyone believe that it's only for our sake that they join the mudslinging of modern democracy? Do the candidates themselves believe this? What an unusual sacrifice that would be.

It's refreshing to work with chimpanzees: They are the honest politicians we all long for. When the political philosopher Thomas Hobbes postulated an insuppressible power drive, he was right on target for both humans and apes. Observing how blatantly chimpanzees jockey for position, one will look in vain for ulterior motives and expedient promises.

I was not prepared for this when, as a young student, I began to follow the dramas among the Arnhem Zoo chimpanzees from an observation window overlooking their island. In those days students were supposed to be antiestablishment, and my shoulder-long hair proved it. We considered power evil and ambition ridiculous. Yet my observations of the apes forced me to open my mind to seeing power relations not as something bad but as something ingrained.

Perhaps inequality was not to be dismissed as simply the product of capitalism. It seemed to go deeper than that. Nowadays this may seem banal, but in the 1970s human behavior was seen as totally flexible; not natural but cultural. If we really wanted to, people believed, we could rid ourselves of archaic tendencies like sexual jealousy, gender roles, material ownership, and yes, the desire to dominate.

Unaware of this revolutionary call, my chimpanzees demonstrated the same archaic tendencies but without a trace of cognitive dissonance. They were jealous, sexist, and possessive, plain and simple.

I didn't know then that I'd be working with them for the rest of my life, or that I would never again have the luxury of sitting on a wooden stool and watching them for thousands of hours. It was the most revelatory time of my life. I became so engrossed that I began trying to imagine what made my apes decide on this or that action. I started dreaming of them at night, and, most significant, I started seeing the people around me in a different light.

I am a born observer. My wife, who does not always tell me what she buys, has learned to live with the fact that I can walk into a room and within seconds pick out anything new or changed, regardless of how small. It could be just a new book inserted between other books, or a new jar in the refrigerator. I do so without any conscious intent.

Similarly, I like to pay attention to human behavior. When picking a seat in a restaurant, I want to face as many tables as possible. I enjoy following the social dynamics—love, tension, boredom, antipathy—around me based on body language, which I consider more informative than the spoken word. Since keeping track of others is something I do automatically, becoming a fly on the wall of an ape colony came naturally to me.

My observations helped me see human behavior in an evolutionary light. By this, I mean not just the Darwinian light one hears so much about, but also the apelike way we scratch our heads if conflicted, or the dejected look we get if a friend pays too much attention to someone else.

At the same time, I began to question what I'd been taught about animals: They just follow instinct; they have no inkling of the future; everything they do is selfish. I couldn't square this with what I was seeing. I lost the ability to generalize about "the chimpanzee" in the same way that no one ever speaks about "the human." The more I watched, the more my judgments began to resemble those we make about other people, such as this person is kind and friendly, and that one is self-centered. No two chimpanzees are the same.

It's impossible to follow what's going on in a chimp community without distinguishing between the actors and trying to understand their goals. Chimpanzee politics, like human politics, is a matter of individual strategies clashing to see who comes out ahead. The literature of biology proved of no help in understanding the social maneuvering because of its aversion to the language of motives. Biologists don't talk about intentions and emotions.

So I turned to Niccolò Machiavelli. During quiet moments of observation, I read from a book published four centuries earlier. *The Prince* put me in the right frame of mind to interpret what I was seeing on the island, though I'm pretty sure that the philosopher himself never envisioned this particular application of his work.

Among chimpanzees, hierarchy permeates everything. When we bring two females inside the building—as we often do for testing—and have them work on the same task, one will be ready to go while the other hangs back. The second female barely dares to take rewards, and won't touch the puzzle box, computer, or whatever else we're using in the experiment. She may be just as eager as the other, but defers to her "superior." There is no tension or hostility, and out in the group they may be the best of friends. One female simply dominates the other.

In the Arnhem colony, the alpha female, Mama, did occasionally underline her position with fierce attacks on other females, but she was generally respected without contest. Mama's best friend, Kuif, shared her power, but this was nothing like a male coalition. Females rise to the top because everyone recognizes them as leader, which means there is little to fight over. Inasmuch as status is largely an issue of personality and age, Mama did not need Kuif. Kuif shared in but did not contribute to, Mama's power.

Among the males, in contrast, power is always up for grabs. It's not conferred on the basis of age or any other trait but has to be fought for and jealously defended in the face of contenders. If males form coalitions, it's because they need each other. Status is determined by who can beat whom, not just on an individual basis but in the group as a whole.

It does not do a male any good if he can physically defeat his rival, if each time he tries to do so the whole group jumps on top of him. In order to rule, a male needs both physical strength and buddies who will help him out when a fight gets too hot. When Nikkie was alpha, Yeroen's assistance was crucial. Not only did Nikkie need the old male's help to keep a powerful third male in check, but he was also unpopular with the females. It was not unusual for females to band together against him. Yeroen, being highly respected, could stop such mass discontent by positioning himself between Nikkie and the screaming females.

But with complex strategies come miscalculations, as Nikkie showed several years later, when he became so intolerant toward his partner, Yeroen, that he lost his support and immediately dropped in rank. Nikkie had underestimated his dependence on the old fox. This is why we speak of political "skills": It's not so much who you are but what you do. We are exquisitely attuned to power, responding quickly to any new configuration.

If a businessman tries to get a contract with a large corporation, he will be in meeting after meeting with all sorts of people from which a picture emerges of rivalries, loyalties, and jealousies within the corporation he is visiting, such as who wants whose position, who feels excluded by whom, and who is on his way down or out. This picture is at least as valuable as the organizational chart of the company. We simply could not survive without our sensitivity to power dynamics.

Power is all around us, continuously confirmed and contested, and perceived with great accuracy. But social scientists, politicians, and even laypeople treat it like a hot potato. We prefer to cover up underlying motives. Anyone who, like Machiavelli, breaks the spell by calling it like it is, risks his reputation. No one wants to be called "Machiavellian," even though most of us are.

About the Contributors

Paul Bloom is a professor of psychology at Yale University. His research explores how children and adults make sense of the world, and his interests include language, social reasoning, art, and morality. Dr. Bloom has won numerous awards for his research and his teaching, and he is the author of many articles and books, including *How Children Learn the Meanings of Words*, and, most recently, *Descartes' Baby: How the Science of Child Development Explains What Makes Us Human.* He lives in New Haven, Connecticut, with his wife, Karen Wynn, and their two sons, Max and Zachary.

He writes: "Religion can be a taboo topic, and it is easy to become annoyed at a psychologist asking why people believe in God, souls, and miracles. But I have long believed that one can learn a lot about the most interesting and intimate aspects of human nature by studying child development. I wrote the article to make the case that this strategy can help us understand the nature and origins of religious belief. This is a fascinating topic from a purely intellectual perspective but it also has social implications. A better understanding of the psychology of religion can help us make sense of contemporary clashes between the sacred and the secular, such as the ongoing debate over the teaching of Intelligent Design and evolution."

Kenneth Chang is a reporter covering the physical sciences for the *New York Times.* In 1994, after seven years as a graduate student in physics at the University of Illinois, he left without his doctorate and went to the science writing program at the University of California, Santa Cruz, to begin a writing career. After three cross-country moves and jobs at the *Los Angeles Times*; *Greenwich Time* in Connecticut; the *Star-Ledger* in Newark, New Jersey; and ABCNEWS.com; he joined the *Times* in 2000. He lives in Jersey City, New Jersey, with his wife and two children.

"Reporting this story was easy," he notes. "Turn on the recorder and listen to Mike Brown, who then tells wonderful stories. The harder part was condensing the tales to fit within the available space. I hoped to capture the fun and serendipity that Dr. Brown finds in science, that it's not an esoteric endeavor, but a search for patterns in something as simple as his newborn daughter's eating and sleeping patterns."

Michael Chorost was born deaf in 1964 due to an epidemic of rubella, and didn't learn to talk until he got hearing aids at age three and a half. He earned a bachelor's degree in English at Brown in 1987 and a doctorate in educational technology at UT-Austin in 2000. He worked between 1999 and 2004 as a technical writer, and now teaches at the University of San Francisco. His memoir of going completely deaf and getting his hearing back with a cochlear implant, *Rebuilt: How Becoming Part Computer Made Me More Human*, was published by Houghton Mifflin in 2005. The book ranges through engineering, computer science, SF, and literature to explore the idea of the "cyborg": the creative fusion of human and computer. Dr. Chorost's Web site is http://www.michaelchorost.com.

Of "My Bionic Quest for *Boléro*," he writes: "This article originated in a chapter on music that didn't make it into my memoir of getting a cochlear implant, *Rebuilt.* It didn't fit into the chronological narrative of the story, so my editor and I reluctantly took it out. The chapter changed dramatically in the process of becoming a *Wired* article, especially because the stochastic-resonance and virtual-channels software had just come out and could be worked into the narrative. Now, in early 2006, I'm using an updated beta version of the virtual-channels software on a processor with four times as much RAM, and am happy to report that my scores in melody-recognition tests have quadrupled. I'm also planning to have my other ear implanted soon, so I can hear in stereo for the first time in 20 years . . . but that's a story for another article."

W. WAYT GIBBS is senior writer and a member of the Board of Editors at *Scientific American* magazine, which he joined in 1992 after a brief stint at *The Economist* in London. Gibbs's work received honorable mention by the NASW Evert Clark/Seth Payne Award for Young Science Journalists in 1995 and 1998. In 1999 Gibbs was selected for a year-long Knight Science Journalism Fellowship at the Massachusetts Institute of Technology. He was awarded both the 2004 Wistar Science Journalism Award and the 2005 AAAS Science Journalism Award for his two-part series on "The Unseen Genome."

"Science writing is always a process of personal discovery," he explains. "Once every while, a story turns you 180 degrees from the direction you set out. For this piece, my initial assignment was to debunk Paul Campos's book *The Obesity Myth* (since retitled *The Diet Myth*). Clearly, my editor and I assumed, Campos must have misperceived (or misrepresented) the firm scientific foundation beneath the 'fact' that the growing girth of Western populations is exacting an enormous toll in death and disability. But as I used Campos's footnotes as a springboard into the primary literature and began hearing leading obesity researchers acknowledge that a great deal of data contradicts the conventional wisdom—what had been my wisdom—I was forced to rethink what it was that needed debunking. Rarely has a story taught me that so much of what I 'knew' was wrong."

GARDINER HARRIS is a reporter in the *New York Times*' Washington Bureau, where he covers public health. He joined the *Times* in April

2003. Previously, he was a reporter for the *Wall Street Journal*; the Louisville *Courier-Journal*; and the San Luis Obispo, California, *Tribune.* He is a recipient of the George Polk and the Worth Bingham journalism awards. He is a Yale graduate.

JACK HITT is a contributing writer for the *New York Times Magazine* and *Harper's*, as well as the public radio program, *This American Life.*

"Archaeology and paleoanthropology are, to be honest, very soft sciences," he remarks. "I've always marveled at the grids at any dig—rectangles of string and prickly concern with Euclidean frames and measures. The need to impose a sense of exacting scientific solemnity has always been paramount since neither discipline can replicate any experiment or perform double-blind studies. So superredundancies are necessary to make claims of truth. But the past rarely accommodates our needs in that way and so seeping into the sparse strata of evidence comes our unconscious interpretation, typically conjuring a world that appears less like how it was than how we now would like it to have been."

ELIZABETH KOLBERT is a staff writer for *The New Yorker* and author of *Field Notes from a Catastrophe: Man, Nature, and Climate Change* (Bloomsbury, 2006). Her three-part series on global warming, which appeared in *The New Yorker* in the spring of 2005, won the American Association for the Advancement of Science's Magazine Writing Award. She lives in Williamstown, Massachusetts, with her husband, John Kleiner, and their three sons.

"I first heard about Harvey Weiss's work on the lost city of Tell Leilan from a friend of a friend," she says. "I think I called up Harvey that same day; I was just fascinated. Tell Leilan's history illustrates how, in the past, even relatively minor changes in the climate have wreaked havoc on human society. Much larger changes are predicted for the future by climate modelers. When you consider the forecasts for the future in the context of what's been learned about the past, it is very, very sobering."

Charles C. Mann's most recent book is *1491* (Knopf), a history of the Americas before Columbus. He is a correspondent for the *Atlantic Monthly*, *Science*, and *Wired*, and he has covered the intersection of science, technology, and commerce for many newspapers and magazines here and abroad, including *Business 2.0*, *Forbes ASAP*, *Geo* (Germany), the *New York Times* (magazine, op-ed, book review), *Panorama* (Italy), *Paris-Match* (France), *Quark* (Japan), *Smithsonian*, *Der Stern* (Germany), *Technology Review*, the *Washington Post* (magazine, op-ed, book review). In addition to *1491*, he has cowritten four other books: *The Second Creation: Makers of the Revolution in 20th-Century Physics* (1986; rev. ed., 1995); *The Aspirin Wars: Money, Medicine, and 100 Years of Rampant Competition* (1991); *Noah's Choice: The Future of Endangered Species* (1995); and *@ Large: The Strange Case of the Internet's Biggest Invasion* (1998). He has also written for CD-ROMs, HBO, and the television show *Law and Order*, and was the text editorial coordinator for the internationally best-selling photographic projects *Material World* (1994), *Women in the Material World* (1996), and *Hungry Planet* (2005). He is a three-time National Magazine Award Finalist, and he has received writing prizes from the American Bar Association, the American Institute of Physics, the Alfred P. Sloan Foundation, and the Margaret Sanger Foundation.

"After this article was submitted," he remarks, "a federal appeals court ruled that jurisdiction in Anna Nicole Smith's case properly belonged not to the federal district court in California that ruled for her but to the state probate court in Texas that had previously ruled against her. Smith appealed to the Supreme Court—backed by the Bush administration, which wished to argue the jurisdictional question, not the merits of the case. In May 2006 the Supreme Court ruled nine to zero in Smith's favor, sending the case back to the federal appeals court in California, where the Smith-friendly district-court ruling will be considered. No matter who finally wins the lengthy litigation, though, it illustrates a central thesis of the article: that as people live longer, the potential rises for older generations to do things that annoy younger generations—marrying strippers and giving them millions of dollars, for example."

D. T. MAX was born in New York City, and he graduated from Harvard College in 1984 with a degree in comparative literature. He was an editor in book publishing; then the book review editor for the *New York Observer*; and, for the past decade, has been a journalist and essayist, mostly for the *New York Times Magazine*. His first book, *The Family That Couldn't Sleep*, a cultural and scientific study of prion diseases, has just been published by Random House. He lives outside of Washington with his wife; two children; and a rescued beagle, who came to them already named Max.

He writes: " 'Literary Darwinism' really began with an essay I wrote called 'Two Cheers for Darwin' in the early 2000s for the *American Scholar*. In that piece I asked—begged, really—Darwinists to shore up their theory before the skeptical started picking it apart—this was before the recent spate of yahooism on the part of various ignorant school boards. The author doesn't always remember what others do from his writing; what I wound up retaining was a sense of the immense seductiveness of Darwinism. Darwinism is itself Darwinian, tapping into a natural proclivity of our minds for creation stories. But what does that mean?—in the absence of very good cognitive psychological studies, probably not much. Anyway, knowing now that Darwinism was in many ways literary, I was curious whether literature might return the favor. Then along came a small fearless band of academics. . . ."

TOM MUELLER earned his bachelor's degree at Harvard and his doctorate at Oxford, where he was a Rhodes Scholar. He is completing his first novel, a historical thriller about the building and rebuilding of St. Peter's Basilica in the Vatican and the mystery of Peter's tomb. He writes freelance for *The New Yorker*, the *Atlantic Monthly*, the *New York Times*, *National Geographic*, *The New Republic*, *Business Week*, and other publications.

"In 1982," he reports, "I had my first run-in with a chess-playing computer, a cheap department store set that gave me a nasty drubbing. Thereafter, whenever I faced a digital opponent, I had the odd sensation of being in the presence of an intelligent being. Researching the genesis of the latest generation of PC chess programs for 'Your Move,' I understood why, in addition to superhuman calculatory prowess, each program had absorbed the playing style, programming methods, philosophy, and ego of its programmer.

"This blend of silicon and grey matter produces chess that sometimes verges on high art."

ANAHAD O'CONNOR is a reporter for the *New York Times*, based in the newspaper's Westchester Bureau. Besides covering crime and politics for the Metropolitan section, he also writes about health for the weekly science section, *Science Times*. He lives in New York City, and he has a degree in psychology from Yale.

"I first became fascinated with autism in 1999," he explains, "as an undergraduate at Yale and right about the time the controversy over thimerosal was reaching its breaking point. My interest in the topic was so great that I ditched my biology course load and switched over to psychology, eventually doing my senior research project on the neural basis of the disorder while also working with autistic children at the Yale Child Study Center.

"When I landed at the *Times* after college as a science reporter, I couldn't resist delving into the subject once again. I covered it deeply enough to get a sense of the passion on both sides. And much to the dismay of scientists and public health officials, the debate over thimerosal is one that very likely will not die down any time soon."

H. ALLEN ORR, professor of biology at the University of Rochester, is an evolutionary biologist whose research focuses on speciation (how a single species splits in two) and adaptation. He is the author of *Speciation* (with Jerry A. Coyne). Orr has been the recipient of a Guggenheim Fellowship, a David and Lucile Packard Fellowship in Science and Engineering, and an Alfred P. Sloan Foundation Postdoctoral Fellowship. He was awarded the Dobzhansky Prize by the Society for the Study of Evolution. In addition to his scientific work, Orr is a frequent contributor of book reviews and essays to *The New Yorker*, the *New York Review of Books*, and the *Boston Review*.

He writes: "I followed the Intelligent Design movement from its beginnings and, at every step, I was surprised by how this confused amalgam of bad biology and worse philosophy grew more popular. But while the press devoted a good deal of attention to the legal and cultural significance of the movement, it paid far less attention to the actual scientific claims made by the movement's proponents. My piece in *The New Yorker* was an attempt to fairly summarize and calmly critique these claims. In the end, I suspect that the Intelligent Design movement will be remembered as an affront to both good science and sound religion."

DENNIS OVERBYE is a science correspondent for the *New York Times*, and the author of *Einstein in Love* (Penguin, 2001) and *Lonely Hearts of the Cosmos* (Little, Brown, 1999). He lives in Manhattan with his wife, the writer Nancy Wartik, and his daughter, Mira.

"Last year," he notes, "as science fans probably were reminded way too often, was Einstein's year, the hundreth anniversary of his discovery of relativity, among other things. As my homage to this august moment, I thought that it would be interesting to look at the edges of the known, where Einsteinian certitudes might be breaking down. I naturally thought of time, since it was a new insight into the nature of at least measured time, namely that moving clocks would appear to run at different speeds, that provided the final breakthrough to relativity. Might there be a new idea and a new revolution lurking out there?

"I hear often from theoretical physicists, mostly string theorists, that space and time are likely to turn out to be illusions of some sort, but unfortunately that is usually where their thoughts end. I was thrilled to my science-fiction heart, however, to discover that there is a rich argument going on in theoretical physics about time travel. Apparently, to the despair of people who would like the universe to make sense, nobody can quite prove that time travel is impossible.

"Is this a great universe—or what?"

RICHARD PRESTON has written five books, including his Dark Biology trilogy: *The Hot Zone, The Cobra Event,* and *The Demon in the Freezer.* He is a regular contributor to *The New Yorker.* He holds a Ph.D. in English from Princeton University, and he has won numerous awards for his writing, including the American Institute of Physics Award and the National Magazine Award. He is also the only non-physician ever to receive the Centers for Disease Control's Champion of Prevention Award for public health. An asteroid is named "Preston" in his honor. (Asteroid Preston travels on a wild orbit near Mars, and it may slam into Mars or the Earth some day, creating an explosion similar to the one that wiped out the dinosaurs.)

ROBERT R. PROVINE is professor of psychology at the University of Maryland, Baltimore County, and author of the award-winning book *Laughter.* After training in developmental neuroscience with two of its founders, Viktor Hamburger and Nobel laureate Rita Levi-Montalcini, he studied development and evolution in species ranging from cockroaches to penguins, using techniques as varied as tissue culture and electrophysiology. Then came the revelation that related neurobehavioral analyses could be conducted with human yawning and laughing. A second revelation was that such research permitted the escape from his windowless laboratory and the opportunity to commune with other members of his species.

"The extraordinary is often hidden in plain sight," he explains. "Common but neglected behaviors such as yawning can be used to strip away learning and culture, and expose our beastly nature—its neurological mechanisms and unconscious motives. Best of all, if you know where to look, cutting-edge science can be done using simple tools, self-experimentation and observations of ongoing behavior.

This is research of the most democratic kind, open to easy confirmation and extension by anyone interested in the scientific enterprise."

David Quammen lives in Montana, and he travels on assignment, most often to jungles, deserts, and swamps. He's a three-time recipient of the National Magazine Award. His work has appeared in *Harper's*, *The Atlantic*, *National Geographic*, *National Geographic Adventure*, *Outside*, the *New York Times Book Review*, and other magazines. His books include *The Song of the Dodo*; *Monster of God*; and, most recently, *The Reluctant Mr. Darwin*.

"I had been keeping a file, for years, on the subject of cloning for its supposed value in rescuing endangered or extinct species," he says. "But the subject didn't become urgent and alive until I spotted that little news clip about Dewey the Duplicate Deer. The hardest thing about writing this story was that I found most of the practitioners in question—Chuck Long, Mark Westhusin, Betsy Dresser, Lou Hawthorne, and especially Duane C. Kraemer himself—enormously likable, yet I had to view their work critically, and with a touch of humor. Maybe they'll forgive me."

Michael Specter has been a staff writer at *The New Yorker* since September 1998, where he mostly focuses on global public health, and on issues of science and technology. Before joining the magazine he was a senior correspondent at the *New York Times*, where he was based in Rome; and, prior to that posting, Moscow bureau chief. Since joining *The New Yorker*, he has written about the AIDS epidemic in India, Africa, Russia, and the United States, as well as on such topics as genetics, malaria, and the Bush administration's approach to science policy. Specter lives in New York City. An archive of his stories can be found at www.michaelspecter.com.

"I started looking into the bird flu somewhat by accident," he says. "In 2002, when SARS suddenly appeared, I decided to report on what seemed to be an emerging epidemic of a deadly new disease. It became clear quickly, however, that experts in the field were almost relieved by SARS, which, while serious, was not devastating. After the earliest reports came out of China, most infectious disease specialists just assumed the disease was H5N1, spreading like a normal flu virus. It was their worst nightmare come true: a pandemic of the bird flu. So I decided to skip SARS and write about that."

NEIL SWIDEY is a staff writer for the *Boston Globe Magazine*. His story, "The Self-Destruction of an M.D.," was included in the 2005 edition of *The Best American Crime Writing*, and his magazine writing has won national headliner awards two years in a row. During his seven years with the *Globe*, his assignments have included immersing himself in the love lives of seniors in retirement communities, spending a season with an all-black high school basketball team in one of Boston's historically all-white neighborhoods, and covering the Arab world during the run-up to the Iraq war. He lives in the Boston area with his wife and three daughters.

"I confess I was bracing for an assault," he writes. "Few topics bring out the dueling advocates as reliably as the roots of homosexuality. So I wasn't surprised when the initial publication of this piece prompted hundreds of letters from readers. I was taken aback, however, by the overall tone of the responses. Liberal gay men and conservative Christian heterosexuals alike said they appreciated the balance and the breadth of the report. Many said they particularly liked my admission that the months I spent marinating in all the available research left me with as many questions as answers. For those of us who work in long-form journalism, there's a natural impulse to want to paper over the holes in our narratives so we don't diminish their power. This story offered a fresh reminder that sometimes there is real power in admitting what you don't know."

FRANS B. M. DE WAAL, Ph.D., is a Dutch-born zoologist known for his work on the social intelligence of primates. His popular books, translated into more than a dozen languages, have made him one of the world's most visible primatologists. De Waal is C. H. Candler Professor in the Psychology department of Emory University and director of the Living Links Center at the Yerkes National Primate Center in Atlanta, Georgia. He has been elected to the National Academy of Sciences (U.S.), and the Royal Dutch Academy of Sciences.

He writes: "This piece comes from my latest book, *Our Inner Ape* (Riverhead, 2005), based on observations that I made as a graduate student. In those days, in the 1970s, comparison between humans and other primates were highly controversial. This has changed for the better, but many people are still unsettled by the idea that we carry a lot of primate heritage around. In my writing, I insist on this message, and try to bring it with some humor to soften the blow."

JONATHAN WEINER majored in English at Harvard. He learned to write about science in the early 1980s while working at the magazine *The Sciences.* In 1985 he left the magazine to write his first book, *Planet Earth*, the companion volume to a seven-part PBS television series. He spent the next twenty years as an independent writer. In 2005 he joined the faculty of Columbia University's Graduate School of Journalism. His books include *The Beak of the Finch*, winner of both the Pulitzer Prize for General Nonfiction and the Los Angeles Times Book Prize for Science; *Time, Love, Memory*, winner of the National Book Critics Circle Award for General Nonfiction, and finalist for the Aventis Science Prize; and *His Brother's Keeper*, finalist for Los Angeles Times Book Prize. He has written for *The New Yorker*, the *New York Times Magazine*, *The New Republic*, and many other newspapers and magazines. In 2000 and 2001 he served as Rockefeller University's first writer in residence.

"One year after I wrote 'The Tangle,' " he reports, "I got back in touch with most of the scientists I'd interviewed. I thought they might have settled the BMAA controversy by then, one way or the other. But in the study of rare diseases, one year is not such a long time. Lucie Bruijn, science director and vice president of the ALS Association, told me, 'I wish science moved that fast.'

"Walter Bradley and Deborah Mash of the University of Miami

were still working on BMAA measurements with Paul Cox. They had not yet published their work. Late at night, when I e-mailed Bradley, he replied immediately with just two words: 'Looking good!' A few days later, Deborah Mash replied with six: 'We are working on it hard.'

"From Mount Sinai, the neuropathologist Daniel Perl wrote to say that he remained skeptical as ever. From Guam, John Steele wrote that he had cooled to Cox's ideas: 'I now read little about the 'batty' hypothesis and people no longer speak of it on Guam.'

"A few of the principal scientists in the BMAA controversy had been angered by my story. They thought I'd given too much space and credit to the other side. One of Cox's critics, Chris Shaw, had sent me some of the hottest pieces of mail I've received in twenty-five years of science writing. I decided not to write him for an update. I did write Cox, but got no answer. Though I have written him many times since the article appeared, but he has never answered.

"I hope to see John Steele shortly in New York; he is being honored for his lifetime of research at a benefit dinner for the Progressive Supranuclear Palsy Society. And Dan Perl and I had another lunch at Hanratty's not long ago. Afterward we went back to his laboratory and he showed me slides through the microscope. First he showed me samples from normal brains, then from Chamorros who died in the epidemic on Guam. Parts of their brains looked like fields of dead snakes. 'Tangle, tangle, tangle, tangle, tangle,' Perl said. 'Just tangle debris. You can't find a normal neuron anywhere.'

"The sight of those tangles brought everything back."

Alan Weisman's work, set in the United States, Mexico, Canada, Central and South America, the Caribbean, Antarctica, Europe, and the Middle and Far East, has appeared in *Harpers*, the *New York Times Magazine*, the *Los Angeles Times Magazine*, *Discover*, the *Atlantic Monthly*, *Condé Nast Traveler*, and *Mother Jones*, and has been heard on National Public Radio and Public Radio International. He is the author of five books, including the forthcoming *The World Without Us*, and he is a senior producer at Homelands Productions, a journalism collective that produces independent public radio documentary series. He teaches international journalism at the University of Arizona.

"This essay began with a question posed to me by *Discover* magazine editor Josie Glausiusz," he explains. "Oddly, she'd recalled a sobering 1994 *Harper's* article I wrote from Chernobyl as hopeful, because I'd described how nature was swallowing the disaster's abandoned surroundings. 'What if people left *everywhere*?' she asked. The resulting piece is a glimpse of what became the most complex, revealing assignment of my life, covering vast swathes of space and time, which has now grown into a book titled *The World Without Us*, to be published in 2007 by Thomas Dunne Books/St. Martin's Press."

KAREN WRIGHT has written about science, health, and the environment for twenty years, first as a staff writer for *Nature* and *Scientific American*, then as a freelancer, whose articles appeared in *Discover*, *Science*, the *Los Angeles Times* and the *New York Times Magazine*, and the *New York Times Book Review*. Between 1999 and 2003 she created and authored the monthly *Discover* column "Works in Progress" with the aid of editor Sarah Richardson, who also helped shepherd this selection. She is at work on a novel.

"Little has changed in Permian-extinction science since 'The Day Everything Died' appeared," she reports. "The NASA investigation of Luann Becker's claims got stalled due to funding delays. Doug Erwin published his book *Extinction: How Life on Earth Nearly Ended 250 Million Years Ago* without saying what nearly ended it. But to my mind, resolution was never the point of this story. Instead, I was trying to capture a living example of the messy controversies that have shaped every history of ideas, scientific or otherwise."

Permissions

"Your Move" by Tom Mueller. Copyright © 2005 by Tom Mueller. All rights reserved. Reprinted by kind permission of the author. First published in *The New Yorker.*

"My Bionic Quest for *Boléro*" by Michael Chorost. Copyright © 2005 by Michael Chorost. All rights reserved. Reprinted by kind permission of the author. First published in *Wired.*

"Earth Without People" by Alan Weisman. Copyright © 2005 by Alan Weisman. All rights reserved. Reprinted by kind permission of the author. First published in *Discover.*

"The Curse of Akkad" by Elizabeth Kolbert. Copyright © 2005 by Elizabeth Kolbert. All rights reserved. Reprinted by kind permission of the author. First published in *The New Yorker.*

"Remembrance of Things Future: The Mystery of Time" by Dennis Overbye. Copyright © 2005 by The New York Times Co. Reprinted with permission.

"Obesity: An Overblown Epidemic?" by W. Wayt Gibbs. Reprinted with permission. Copyright © 2005 by Scientific American, Inc. All rights reserved.

A Note from the Series Editor

Submissions for next year's volume can be sent to:

Jesse Cohen
c/o Editor
The Best American Science Writing 2007
HarperCollins Publishers
10 E. 53rd Street
New York, NY 10022

Please include a brief cover letter; manuscripts will not be returned. Submissions made electronically are also welcomed and can be e-mailed to jesseicohen@netscape.net.